Henry Fairfield Osborn

Palaeontological Report of the Princeton Scientific Expedition of

1877

Henry Fairfield Osborn

Palaeontological Report of the Princeton Scientific Expedition of 1877

ISBN/EAN: 9783337329266

Printed in Europe, USA, Canada, Australia, Japan

Cover: Foto ©berggeist007 / pixelio.de

More available books at **www.hansebooks.com**

CONTRIBUTIONS FROM THE MUSEUM OF GEOLOGY AND ARCHÆOLOGY

To ....

*From the E. M. Museum of Geology
and Archæology of the College of
New Jersey.*
                    A. GUYOT, Director.
Princeton, N. J.

OF

The Princeton Scientific Expedition

OF 1877.

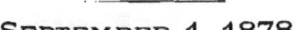

BY
HENRY F. OSBORN,
WM. B. SCOTT,
FRANCIS SPEIR, Jr.

SEPTEMBER 1, 1878.

NEW YORK:
S. W. GREEN, PRINTER, Nos. 16 AND 18 JACOB STREET.
1878.

CONTRIBUTIONS FROM THE MUSEUM OF GEOLOGY AND ARCHÆOLOGY
OF PRINCETON COLLEGE.

No. 1.

PALÆONTOLOGICAL REPORT

OF

The Princeton Scientific Expedition

OF 1877.

BY
HENRY F. OSBORN,
WM. B. SCOTT,
FRANCIS SPEIR, Jr.

SEPTEMBER 1, 1878.

NEW YORK:
S. W. GREEN, PRINTER, Nos. 16 AND 18 JACOB STREET.
1878.

Palæontological Division.

PROF. JOSEPH KARGÈ,
ROLLIN H. LYNDE,
HENRY F. OSBORN,
JOTHAM POTTER,
WM. B. SCOTT,
FRANCIS SPEIR, Jr.

TO THE HONORABLE THE PRESIDENT AND BOARD OF TRUSTEES OF THE COLLEGE OF NEW JERSEY:

Gentlemen:

I have the honor to transmit herewith the Palæontological Report of the College Scientific Expedition of 1877.

The fossils collected by the Palæontological party, and deposited by the chief of the expedition in the Geological Museum, consisted of two sets, one numbering some two thousand specimens of fossil plants and insects from the tertiary beds of Central Colorado, the other of a considerable series of fossil vertebrates, mostly mammals, from the tertiary beds of Wyoming Territory, around Fort Bridger. This last collection has been studied and worked out with unabated zeal and diligence by the three post-graduate members of the Palæontological party, Messrs. H. Osborn, W. Scott, and F. Speir, who devoted most of the time of their course to this special work, with what success this Report will show.

It will be a source of gratification to the generous friends of the College, who furnished means for the Scientific Expedition of 1877, that it not only enriched our Museum to so great an extent, but did more still by fostering in our College a thorough study of Palæontology, which could not have been undertaken without such means as these thus placed at the disposal of our students.

The fossil insects and plants have been intrusted to the hands of the best specialists for determination. Dr. S. Scudder, of Cambridge, Mass., has kindly consented to revise the insects, Prof. G. L. Lequereux the plants.

Very Respectfully,

A. GUYOT,
Director of the E. M. Museum of Geology and Archæology.

PRINCETON, June 1, 1878.

SIR: We transmit herewith our report upon the Palæontological collections made by the Princeton party in the summer of 1877.

The following persons constituted the Palæontological division: ROLLIN H. LYNDE, HENRY F. OSBORN, JOTHAM POTTER, WM. B. SCOTT, FRANCIS SPEIR, Jr. The division remained in Colorado from the first of July until the first of August, when, under the direction of Professor Kargè, it left the main party, and passed the month of August in Wyoming, returning in the first part of September. The Colorado collections were mostly made in the (probable) Miocene beds near Florissant, and in the beds near the Garden of the Gods, variously referred to the *Dakota* and *Wealden* groups. In Wyoming, with Fort Bridger as a base of explorations, the division was wholly occupied in the Bridger series, camping successively on Smith's Fork, Henry's Fork, and Dry Creek, and exploring the beds adjacent.

It has been our endeavor, in confining our attention to the remains of vertebrated animals collected during the trip, not merely to catalogue the direct results, but, by the aid of fresh materials, to supplement the work of others. For, of the descriptions and data of the Bridger Eocene Fauna published up to the present time, we find that even those which have been most accurately prepared are lacking in important details; and that, owing to imperfect materials, large gaps yet remain in our knowledge of genera and species named and classed years ago. Although such supplementary work may appear, at first sight, tedious and ill-directed, we are confident that in the end it will prove of some value to science, and that it is therefore well worthy of our effort. While our field work did not extend beyond a region previously well explored, we obtained material by means of which we are able to add a number of new ossils to the Eocene Fauna of the Bridger group.

In the preparation of this report we have experienced much difficulty in assigning some of our specimens to their proper genera and species. For, while we have desired to respect the classifications made by others, we have in many cases found it impossible to do so, owing to uncharacteristic definition, which, without doubt, has been unavoidable. In all cases of uncertainty, we have adopted the classification which appeared to be the best established. This, in short, has proved the only available course.

The drawings have been executed with much care as to accuracy of proportion and outline. They are, with one exception, the work of a member of the party; and they are inserted simply to illustrate certain parts of the context, which would be unsatisfactory without reference to figures of the kind.

Now that the present work is ready for the press, we are very sensible that it must contain errors which, while they have escaped our notice, will be readily detected by eyes more experienced. These, we trust, will be excused, when it is remembered that we are just entering a field which others have explored for years; and opening a work which Princeton, with her many other lines of study, has never hitherto attempted.

We take this opportunity to return our most hearty thanks to General Flint, to Judge and Dr. Carter, to Mr. Hamilton, and other officers and residents at Fort Bridger, who, by their good will and liberal assistance, contributed largely to our success. Our gratitude is also due to Professors Leidy and Cope for their generous aid, both in the way of advice and of material put in our hands for comparison.

The following pages do not embrace descriptions of the entire collections made by the Princeton party last summer. The valuable specimens of fossil plants and insects have passed into other hands.

Respectfully submitted,

HENRY F. OSBORN,
WM. B. SCOTT,
FRANCIS SPEIR, Jr.

Dr. Arnold Guyot,
Director of the E. M. Museum.

INTRODUCTORY NOTE

UPON THE

GEOLOGY OF THE BRIDGER BASIN.

FORT BRIDGER is a government military post, situated on the high southern plateau of western Wyoming Territory, in the midst of one of the most interesting geological regions of the world.

The country on all sides was once the bottom of a great eocene lake, the water of which was probably slightly brackish. Whether this lake district had direct communication with the ocean, is undetermined as yet, but there is a possibility that it had.*

The tributaries of the Green River, which drain this plateau, render the valleys along the edges of the streams green and wooded. Beyond this fertile strip, wide, barren plains extend, covered by a dense growth of short "sage brush" (Artemisia tridentata).

From the fragmentary débris lying scattered over the surface of the ground, it would seem as if the various streams formerly were of much greater size and volume than they now are, and that long after the eocene lakes had been drained rivers of considerable size ploughed up the lake bottoms, excavating an immense area. The formation known as "Mauvaises Terres" rises abruptly from the valleys, and extends in a series of plateaus, one above the other, on either side.

The high land shows the effect of violent erosion in two forms ; first, the irregular and jagged cones that appear upon the sides of the high benches ; and, second, the isolated butte structure, rising directly out of the plain.

The bad lands of Cottonwood Creek, Henry's Fork, Dry Creek, etc., are examples of the first, and Bridger Butte the best known example of the second.

Bridger Butte, six miles to the south-west of Fort Bridger, rises to a height of over a hundred feet, and is about two miles long ; its sides slope steeply up, and its level top serves as a landmark that can be seen miles away.

The stratification throughout this whole formation is nearly horizontal, and across the valley can be distinctly noted, owing to the color and appearance of the various layers.

No satisfactory explanation has been given of the causes which occasioned the removal of the waters of these lakes, nor of the agencies necessary to ac-

---

* Ichthyic fauna of the Green River shales, Hayden's Surveys, vol. iii., No. 4, p. 819.

complish the great excavations that now show the former bottoms of the basins. These and like points future investigation will undoubtedly solve.

It is certain that the level of the lakes varied at different times, and also that great stretches of marshy land surrounded their borders.

The first fact is proved by the characters of the different layers of strata; the second by the fossil remains entombed. A careful study of the formation of the beds of Cottonwood Creek, at a point about fourteen miles south of Fort Bridger, yielded the following result, which will serve as an example illustrative of the regular formation in this section.

Three distinct lines of bluffs are to be noticed, the first rising to a height of one hundred and fifty feet; from these extends a plain, gently sloping south-west to the foot of the second line of cliffs; these seem high, owing to the downward slope of the plain, but they really rise only fifty feet higher than the first.

On the top of the second, but less broad than the first, extends a level plain, with a slight dip to the south-east; at the end of this, the third line rises two hundred feet above the top of the second line of bluffs, making a total height, in the series, of four hundred feet above the level of Smith's Fork.

The strata throughout are nearly horizontal, and are of different color and composition.

Specimens of the rocks and clays from this section were gathered, and submitted for analysis to Professor Cornwall, of the Scientific School at Princeton, who has kindly furnished us with the following notes:

No. 1, very friable, light greenish-gray sedimentary rocks, consisting chiefly of crystalline grains of quartz, orthoclase, and hornblende (this often in slender crystals), with a little dark mica, and irregular fragments of a light-greenish, transparent, not dichroitic mineral. The above are not perceptibly affected by hot hydrochloric acid. The whole is loosely cemented with a calcareous clay, containing considerable phosphoric acid. These rocks might result from the disintegration of a neighboring hornblendic granite.

The greatest mass of the strata is made up of this kind of rock, and it is this which gives the peculiar color to the "Mauvaises Terres."

Above this in places is found a second kind, which is a light gray indurated clay, with a slight greenish tint. It contains much fine crystalline quartz, with considerable carbonate of lime, and a little phosphoric acid.

It appears to be of similar origin with the first, but was deposited in quieter waters. This mineralogical evidence is strengthened by the fact that no remains of mammals were found in strata of this kind, but only shells regularly deposited in layers one above another.

The third kind is found in thin layers, overtopping the highest line of buttes; it consists of very fine-grained dark-brown sandstones, containing a considerable proportion of carbonate and phosphate of lime. They are hard and tough, and are mechanically deposited, and no fossils are found in them.

In No. 2 the indurated clay is often found above a coarser sandstone than No. 1, but of the same general appearance; with the exception that it contains smooth, rounded pebbles, which were deposited either on a beach or in running water. In this stratum the fossils found are separate bones, often showing marks of having been broken before they were silicified. This would prove that the lake level was changing continually.

Several skeletons of animals have been found in a standing position, with their legs slightly stretched out. They were probably mired, and, being unable to extricate themselves, died in an erect position. This fact affords evidence that extensive swamps surrounded the borders of the lake.

The snow-water and the spring rains wear deep gulleys through the lines of cliffs, and wash down fragments of bones into the dried-up water-courses. These pieces can generally be traced up to the spot from which they came, and the rest of the skeleton can so be secured.

The fossils found in the eocene of Wyoming are entirely petrified, presenting a darker appearance than their matrix; but loose bones washed out, and subjected to the influence of the sun and rain, often become bleached so as to resemble in color modern bones.

The state of preservation of the fossils differs according to the matrix in which they are found. Generally speaking, the remains found in the lower lines of buttes have been considerably distorted by pressure; while those from the highest line have suffered very little from this cause.

MAMMALIA.

PRIMATES.

TOMITHERIUM, Cope.

Ann. Rep. U. S. Geol. Survey of Terrs. 1872, p. 546.

"DENTAL formula of the inferior series: I. 2, C. 2, Pm. 4, M. 3. The last molar has an expanded heel. The third premolar consists of a cone with posterior heel. Fourth premolar exhibits, besides its principal cone, an interior lateral one and a large heel. The true molars support two anterior tubercles, of which the inner is represented by two distinct cusps in one or more of them, and the external is crescentoid in section. The posterior part of the crown is wide and concave, and bordered at its posterior angles by an obsolete tubercle on the inner, and an elevated angle on the outer side." (Palæontology, Wheeler's Survey, IV (pt. 2), p. 135.)

TOMITHERIUM ROSTRATUM, Cope.

Loc. cit., p. 548.

This species exhibits considerable variation, both in size and proportion of the teeth. We have two specimens of it, one of which agrees exactly with the measurements given by Professor Cope, while the other is stouter, and probably belonged to an old male.

The incisors are too much broken for description. The canine has a long, stout fang, which is subcircular in section;

the crown is compressed, and shows a distinct cutting edge posteriorly. It is short, and tapers rapidly; in this respect differing from *Notharctus*, Leidy, which has a long recurved canine. The first and second premolars are inserted each by a single fang; the third and fourth by two. The third premolar consists of a simple conical crown with a small posterior heel; and the fourth has this heel enlarged, with a small tubercle developed inside and slightly behind the principal lobe. All the premolars have striated enamel, and very feebly marked basal ridges.

The true molars are considerably larger than the premolars; the third is the longest of the series, and the second is the widest. They all seem to be inserted by two fangs. The posterior fang of the last molar is a flat quadrate, of the same size throughout, having a great fore-and-aft diameter, and apparently no nerve cavity. The mandible is strong but shallow, has a curved alveolus and lower margin, and the teeth form a curve with convexity outwards. The symphysis is short and oblique.

This genus presents a close resemblance to the modern *Lemur*, but at the same time shows several differences. We may give these differences in systematic order: (1) Greater number of premolars, in *Tomitherium* = 4, in *Lemur* = 2. It will be observed, however, that the first and second premolars of *Tomitherium* are very small and single-rooted, and that their disappearance is a comparatively slight change. (2) The canines are subcircular in section, not nearly so much compressed. (3) Greater breadth of the molars in proportion to their length. (4) Two internal cusps on the molar. (5) Much greater size of the last molar. (6) Greater depth and thickness of the jaw. (7) Greater curvature of alveolus and lower margin of ramus. (8) Longer and more oblique symphysis.

The third and fourth premolars of *Tomitherium* correspond almost exactly to the first and second of *Lemur*, but they are not quite so high and sharp. The interior tubercle of the second premolar is not so distinct in the latter genus.

In the second specimen the cusps are all low, and the crests prominent, giving the molars something of the appearance of *Opisthotomus*. This difference is probably sexual.

Measurements.

	M. T. No. 1.	M. T. No. 2.	M. Lemur.
Length of entire molar series............	·038	....	·032
Length of premolar series.............	·017	....	·0125
Length of true molar series............	·021	....	·0195
Fore-and-aft diameter of canine.........	·003	·005	·0065
*Length of last molar................	·008	....	·005
Width of last molar..................	·004	....	·003
Length of second molar...............	·007	·007	·0072
Width of second molar...............	·005	·006	·005
Length of last premolar.............	·0055	...	·007
Length of penultimate premolar.........	·004	....	·005
Depth of jaw at second molar..........	·012	·012	·0095
Thickness of jaw at second molar........	·0065	·009	·004

Specimen No. 1 was found at Cottonwood Creek, and specimen No. 2 at Henry's Fork, Wyoming.

HYOPSODUS, Leidy.

Pr. Ac. Nat. Sc., 1870, p. 110.

Lower teeth : I. 3, C. 1, Pm. 4, M. 3, in uninterrupted succession. Last molar has cusps in opposing pairs ; the antero-internal cusp on all the molars is single ; the last molar has a heel, and the last premolar has an inner cusp. The true molar cusps are all high and simple.

HYOPSODUS PAULUS, Leidy.

Loc. cit., p. 110.

This is one of the most common fossils found in the Bridger Basin. We have numerous specimens, chiefly from Henry's Fork, exhibiting a large range of individual and sexual variation. As Dr. Leidy has already pointed out, the strength and depth of the lower jaw is extremely variable, increasing with the age of the animal ; so that the most worn teeth are associated with the deepest jaws.

* In measurements of teeth we use the word *length* to mean antero-posterior diameter, and *width* to mean transverse diameter.

In addition to the jaws and teeth, (which have been very accurately described), we have a portion of a pelvis and femur, which are important as tending to confirm the reference of this genus to the Lemurs.

The *pelvis* is represented by the acetabulum and a small portion of ilium and ischium. They resemble the corresponding parts in the skeleton of *Stenops gracilis*. The acetabulum is a long oval, not subcircular, deep, and quite narrow from side to side. It appears to be directed nearly straight outwards. The ilium is narrow, has a concave gluteal surface, and a prominent acetabular border. The pubis evidently projected forward, making a right angle with the ilium; while the ischium is slender and nearly in the same plane with the ilium. The *femur* is thoroughly lemurine in shape. The shaft, (as much of it as is preserved), is straight and subcylindrical; it is not flattened even distally, but becomes very thick and trihedral in shape just above the trochlea. The trochlea is long, and rises obliquely upon the shaft; the groove is deep, and the two divisions are asymmetrical, the external somewhat the larger. The condyles are large. They are but slightly convex in either direction and project backwards, and are broadest posteriorly. The internal is the larger. They are separated by a deep but not wide popliteal groove, which does not extend into a popliteal fossa. The position and shape of the condyles are such as show that the femur must have been very oblique to the tibia, as in the other lemurs. The tuberosities, especially the internal, are very large and prominent. The whole distal end has an asymmetrical appearance, owing to the greater size of the internal condyle.

Measurements.

	M.
Breadth of ilium at acetabulum	·0065
Vertical diameter of acetabulum	·007
Transverse diameter of acetabulum	·009
Fore-and-aft diameter of shaft of femur above trochlea	·0065
Transverse diameter of shaft above trochlea	·0065
Width of trochlea	·004
Length of trochlea	·0075
Breadth over condyles	·010
Breadth of inner condyles	·005

OLIGOTOMUS, Cope.

Ann. Rep. U. S. Geol. Survey of the Terrs., 1872, p. 607.

"Molars constructed much as in *Hyopsodus* and *Lophiotherium*, viz., with two external subtrihedral cusps which wear into crescents, the posterior connected by a low oblique ridge with the basis of the anterior cone of the inner side; the latter with two conic cusps. It differs from these genera and from *Orotherium* in the possession of two premolars; the inferior molars are probably six, leaving four true molars."

OLIGOTOMUS CINCTUS, Cope.

Loc. cit.

Represented in our collection by the penultimate lower molar, and a caudal vertebra, which has a remarkably long, slender, and simple centrum, with rudimentary metapophyses

OPISTHOTOMUS, Cope.

Wheeler's Survey, Pal. v. iv., pt. 11, p. 152.

"The inferior lower molars do not display a bifid or double anterior cusp; and the crowns exhibit two anterior cones, and an inner cone and outer crescent posteriorly. The posterior crescent is well defined, and is continued on a narrow crest to the anterior inner tubercle. The posterior molar presents the peculiarity of a series of three cusps in one line, the median having another or lateral cusp near it."

This genus has hitherto been found only in the Wahsatch formation; but we have discovered it to be represented in the Bridger series by the species *O. astutus*, Cope. Our specimen consists of a part of the ramus mandibuli containing a single molar tooth.

CARNIVORA.

SINOPA, Leidy.

A genus of small carnivorous animals, which Dr. Leidy regards as intermediate between the recent *Canis* and the extinct *Hyænodon*. Owing to the fragmentary condition of the remains found, no satisfactory generic definition has been given.

From the portion in our collection, we are able to throw some further light upon the genus, summing up the generic characteristics thus: Small carnivores, which have the last upper premolar as sectorial (thus differing from *Hyænodon*), the other premolars simple and conical.

The sectorial is shorter, antero-posteriorly, than the preceding tooth; has a short blade of a single lobe, and a large cusp developed from the posterior part; a cingulum surrounds the entire crown. The *lower* sectorial has the blade of a single lobe, and with a short heel.

SINOPA RAPAX, Leidy.
Proceedings of Ac. Nat. Sc., 1871, p. 115.

In addition to the molars of the lower jaw, described by Dr. Leidy, we have what corresponds to the third and fourth premolars of the fox, their dental formulæ being probably the same.

The third premolar is small and pointed; differing from the corresponding tooth in the fox, (1) in its being less compressed, (2) in its shorter antero-posterior diameter, (3) in thes traighter and more nearly equal margins, and in (4) the absence of a posterior heel.

The tooth is inserted by two fangs, as in *Canis* and *Hyænodon*. The posterior shows a rudiment of a third, which is connate with its entire length above the alveolus. There is an indistinct cingulum around the entire crown.

The fourth premolar has a very curious shape. The blade of this tooth resembles the crown of the third, but is smaller. It is inserted by three fangs, the disposition of which is opposite to that in *Canis*, the internal, being on the same transverse line as the posterior external, instead of the anterior, as in *Canis*. From the internal fang arises a sharp cusp, which is nearly as large as the blade of the tooth, the two are connate at base. The anterior face of the crown is much worn, and there is a small anterior heel formed by the basal ridge. The cingulum is complete all around.

The *maxillary* does not show the outward bulge at the third premolar, which is so marked in the fox. The alveolus is straighter, and the palatine plates are comparatively thicker and flatter. The infraorbital foramen is oval, and not so much compressed as in the fox, to which it corresponds very nearly in position, though situated slightly forward as in *Hyænodon*.

Measurements.

Upper Jaw.

	M.
Length of third premolar	·007
Breadth of third premolar	·004
Length of fourth premolar	·007
Breadth of fourth premolar	·007

Lower Molars, from Dr. Leidy.

Length of last premolar	·0075
Length of first molar	·009

These exhibit nearly the same proportionate size as in the gray fox.

Genus ——. *Species* ——.

Sacrum (Plate IX., Fig. 8).—This peculiar sacrum is composed of only one true vertebra; there may have been one or more pseudo-sacrals, but this is not certain.

The centrum is very long, strongly depressed, and straight on the inner margin, not curved as in the sacrum of most mammals. The anterior articular face is much depressed, and is one third larger than the posterior. The neural canal is low and subtriangular, resembling very much that of *Canis*. The pleuropophysial plates for articulation with the ilia are large and stout. The laminæ are heavy and concave on their upper side, supporting a very long, stout spine, which is retroverted and decidedly tuberous at the end.

The pedicles are deeply notched behind; and on the fore part, just inside the metapophyses, there is a deep fossa.

The chief features of this sacrum are decidedly carnivorous; but to what genus or family it should be referred we are unable to say.

It has some of the characteristics of *Canis*, but the length and retroversion of the spine, as well as the size of the centrum, prevent this classification. In the general form of the pleuropophysial plates it approximates to the *seals*; while in its angle and curvature, it partakes of the character of the *Ursidæ*.

The chief point of interest in this fossil centres in the fact that it was found only a few feet from the brain cast that is described below.

Measurements of Sacrum.

	M.
Length of centrum	·031
Long diameter of anterior articular face	·024
Long diameter of posterior articular face	·017
Width of neural canal	·019
Height of neural canal	·011
Length of neural spine	·036
Extreme width of sacrum	·052

MEGENCEPHALON.

MEGENCEPHALON PRIMÆVUS. *Gen. et spec. nov.*

In close proximity to the pelvis of the *Uintatherium Leidianum*, in one of the upper beds we found an intracranial cast, separate from the bone which had enclosed it, and in such preservation as to warrant a partial determination, at least, of the type to which it belonged. Wishing to obtain as full information as the nature of the cast permitted, we put it in the hands of Dr. Spitzka, of New York, who kindly undertook an examination, and sent us the following as the result:

"SIR: The specimen submitted to me is the intracranial cast of some species of Placental Mammals. The cranium had been subject to the influences of the atmosphere, etc., for a considerable period preceding the formation of the cast, and therefore the cast reflects the sutural dislocations which occurred in consequence. The base of the brain cast it is not advisable to attempt to expose, on account of the treacherous nature of the material. The convolutions corresponding to the internal aspect of the *Os temporale* have not been clearly demarcated by the bone surface. The two narrow eminences on it are casts of the grooves of the middle meningeal arteries. The convolutions of the occipital surface had been well marked, but somewhat obliterated through denudation, etc. The important region bordering on each side of the median fissure, and corresponding to the fronto-parietal suture, is unfortunately as good as destroyed ; and with this destruction the key to the interpretation of the specimen is lost. However, this much can be stated with absolute certainty, that the frontal region is sufficiently well preserved to state that its convolutions do not correspond to those of the brain of the tapir, rhinoceros, hippopotamus, elephant, pig, horse, hyrax, manatus, or any ruminant or cetacean.

"They also differ in important particulars from those of the *Canidæ*, differ less from those of the *Felidæ*, still less from the *Ursidæ*, although corresponding to none of them. The outline of the cerebral cast is found in two living animals—the marine otter and the seal. But in the seal the gyri show the transverse interrupting series of sulci, characteristic of extreme brachycephaly ; and it therefore cannot belong to any animal corresponding to the seal.

"The sea otter's convolutional details are unknown to me, and I believe have not yet been studied. I therefore content myself with stating that the outline of this cast corresponds to the outline of the sea otter's cranium.

"It would help us a great deal if we could decide the existence or non-existence of a bony tentorium. The sutures of this cranium, as far as I can reconstruct them, ran as in the diagram.

"We may state definitely that this was not an ursine, feline, or canine brain, nor the brain of any terrestrial viverrine. It is an open question between an *aquatic carnivore* and an *aquatic pachyderm ;* and although not placing my conclusion on an exact basis, yet, in view of the general outline, the course of the convolutions, and the course of the sutures, I incline to the former view.

"It certainly corresponds to no known brain of a living creature. In one point I was inclined to suspect it to be a pachyderm, namely, the decided assymmetry of some of the sulci, but this, by itself, is not decisive."

"DR. SPITZKA.

"308 East 123d street."

The interesting letter quoted in full above, contains as near a determination of the character of the animal to which the brain belonged, as the nature of the cast and the materials for comparison would permit. In a later report, by means of more complete comparative material, we hope to be able to reach a more satisfactory conclusion. However, as Dr. Spitzka writes, the *general outline*, the *course* of the *convolutions*, and the *line* of the *sutures* offer strong presumptive evidence that the cast belongs to one of the *Aquatic carnivores*. Not far from the brain was found a sacrum, which is described above as belonging to some carnivore, though further determination was impossible. Whether there was any connection between the two is difficult to state. The presence of an aquatic carnivore in the Bridger eocene is new to science ; but, aside from this, the brain is of a much higher order than previous discoveries would lead us to expect in such an early formation.

Professor Marsh's researches have led him to form the opinion that the eocene mammals had brains of a low character ; but this specimen shows that this is not true of all, if it is of most of them. The convolutions are not only numerous and well marked, but they are complicated, showing the transverse as well as the longitudinal folds. To such an extent is this true that the brain will bear comparison with the very highest modern carnivorous types.

We hope to be able to give further notes upon this interesting specimen at a later date.

PERISSODACTYLA.

ANCHITHERIUM.

Von Meyer, Jahrbuch für Mineralogie, 1844, p. 298.

ANCHITHERIUM—?

A small calcaneum and astragalus of equine type are provisionally referred to this genus until further material enables us to determine them with certainty.

The *astragalus* has narrow and very oblique condyles, which are more equal in size than in *Orohippus;* the neck is very short, the internal condyle reaching to the face for the navicular; the posterior projection of this condyle is much shorter than in that genus. The articular face for the navicular is quadrate in shape and concave; the cuboid face is very narrow. The articulation with the calcaneum is made by a narrow, convex face. When the two are in position the navicular face of the astragalus is in the same horizontal line as the cuboid face of the calcaneum, thus resembling the arrangement of the horse's tarsus rather than that of *Orohippus.*

The *calcaneum* is a short, slender bone, having the upper and lower margins convergent toward the tuberosity, and not parallel as in *Orohippus.* The tuberosity is especially small. The face for the cuboid is very narrow.

From the articular facets of these two bones we can see that the tarsus resembled very much that of the modern horse, with a broad, short navicular, and a narrow cuboid. The strata in which these remains were found were somewhat higher than those containing the bones of *Orohippus.*

Measurements.

	M.
Greatest length of astragalus	0.021
Greatest breadth of ditto	.018
Length of neck of ditto	.005
Width between the condyles	.010
Vertical diameter of face for navicular	.012
Tranverse diameter of ditto	.012
Length of calcaneum	.046
Width of face for cuboid	.006

From Henry's Fork.

OROHIPPUS, Marsh.

Am. Jour. Sc. vol iv., p. 207, third series.

Generic Characteristics.—"The crowns of the upper true molars are composed of a pair of external cusps similar to those of anchitherium. There are two coresponding inner tubercles, from which ridges extend obliquely to the anterior inner margin of the outer cusps; but the anterior ridge is divided so as to form an intermediate anterior tubercle. All the teeth preserved have a distinct basal ridge."

Species Known.—

OROHIPPUS PUMILUS, Marsh.
OROHIPPUS MAJOR, Marsh.
OROHIPPUS AGILIS, Marsh.
OROHIPPUS GRACILIS, Marsh.

OROHIPPUS PUMILUS? Marsh.

Specimen obtained. Penultimate and third superior molars, with part of zygoma.

From Cottonwood Creek.

OROHIPPUS MAJOR? Marsh.

Femur (Plate IX., Fig. 1).—The femur has a small, nearly hemispherical head, developed on a long and slender neck; the head is but slightly out of the axis of the shaft, and has a large pit for ligamentous insertion. The shaft is long, simple, and curved slightly forward. At the proximal end it is broad and flattened axially; below this it becomes expanded fore and aft, but it thickens greatly at the distal end, just above the condyles. The great trochanter is large and retroverted, rising above the head, with two prominences rising from it, one on top, the other back. The digital fossa is wide and deep, penetrating far into the great trochanter. The second trochanter is a small rounded ridge ; the third trochanter is large and prominent, curving slightly forward. The condyles are long and narrow, projecting very far back, and are separated by a wide and deep popliteal groove. From the external condyle a low ridge runs obliquely, forming the upper border of the shallow popliteal space. The trochleæ are long, very convex, deeply grooved, and symmetrical.

Measurements of Femur.

	M.
Width between head and great trochanter	·019
Width at third trochanter	·034
Diameter of head	·019
Width at condyles	·028
Width of trochlea	·015
Height of great trochanter	·015
Diameter fore-and-aft of shaft at middle	·018

The *tibia* (Plate IX. Fig., 3) is very long and heavy, with broad proximal articular face, the inner borders of which are prolonged upward and separated by a groove.

The shaft at the tuberosity is subtriangular, with strongly concave sides. The tuberosity is prominent, with a deep pit on its upper surface for the insertion of the ligament of the patella. The shaft below becomes sub-cylindrical, and decreases regularly in size downward. Its curvature is forward. The distal articular face is divided by a smooth ridge into two deep facets. The malleolus is long.

The *fibula* (Plate IX., Fig. 2) is distinct, straight, and very slender. The distal end is but slightly expanded, and is strongly marked by a vertical groove externally. The proximal end articulates with the overhanging portion of the proximal face of the tibia.

Measurements of Tibia.

TIBIA.

	M.
Length	·178
Width of proximal surfaces (transverse)	·032

Width of proximal surfaces (antero-posterior)........................... ·019
Transverse diameter of shaft... ·022
Antero-posterior diameter of distal articulation....................... ·018
Transverse diameter of distal articulation............................. ·018

The tarsus.—The *astragalus* (Plate IX., Fig. 5) has the condyles asymmetrical and divided by a deep groove; the head is narrow, with the neck elongate. The face for the cuboid is small, and confined to the external border.

The *calcaneum* (Plate IX., Fig. 4) is long and compressed, with its upper and lower margins straight and nearly parallel; its tuberosity is large. The face for the cuboid is small.

The *navicular* (Plate IX., Fig. 6) is proportionately longer and narrower than it is in the modern horse. The internal and middle cuneiforms were probably separate.

Metatarsals (Plate IX., Fig. 7).—Three in number. Are very much shorter proportionately than in the modern horse. In shape they are compressed and arched forward. The distal ends are flattened vertically, arched forward, and deeply grooved in the middle.

The *phalanges* (Plate IX., Fig. 7) are very short, rather stout, and very smooth and convex above.

The ungual phalanges are very thin and crescent shaped.

Measurements.

ASTRAGALUS.

	M.
Greatest width	·029
Greatest length	·022
Length of navicular facet	·017
Width of navicular facet	·012
Length of tibial trochleæ externally	·016

CALCANEUM.

	M.
Total length	·058
Total width	·019
Depth in front	·022
Length of heel	·035
Depth of heel	·018
Length of cuboid facet	·014

NAVICULAR.

	M.
Width	·019
Length	·010

PHALANGES.

	M.
Length of first phalanx	.022
Width of first phalanx	.012
Length of second phalanx	.013
Width of second phalanx	.010
Length of ungual phalanx	.005
Width of ungual phalanx	.010

PALÆOSYOPS, Leidy.

Hayden's Geological Survey of Montana, 1871.—Proceedings Academy Natural Sciences, Philadelphia, 1871, p. 118.—*Limnohyus.*—Marsh, American Journal Science and Arts, 1872, p. 124.

Generic characters.— The dentition is full, I. 3, C. 1, Pm. 4, M. 3; the same in lower jaw. The internal cones of the superior molars isolated from the crescentoid crests. One inner tubercle on the last three premolars. One internal cone on the last superior molar. In lower jaw, true molars with four acute tubercles alternating in pairs and connected by oblique crests. The last molar adds a fifth posterior tubercle. The last premolar lacks the posterior inner tubercle. The canines are in continuity with the incisors.

A broad, triangular forehead. A wide zygoma. Long, projecting nasals. Large temporal fossæ. High sagittal crest. Prominent and nearly vertical occiput.

PALÆOSYOPS MAJOR, Leidy.

Survey of Wyoming, 1871, p. 359.—*Limnohyus robustus.*—Marsh, American Journal Science and Arts, 1872, p. 124.

Specific characters.—Sagittal crest short and thick. Temporal fossæ not very deep. Frontals diverge rapidly. Occipital condyles wide and low; the same is true of the foramen magnum. Meatus auditorius high and deep. Glenoid cavity shallow; no internal process. As compared with *P. Paludosus*, post-glenoid process not so much compressed. Occipital region comparatively higher and not so concave. The occipital condyles are more prominent. The zygomas are lighter. Head larger than *Paludosus*, perhaps not so large as *Vallidens*.

Description from (1) a head, complete, but distorted, with complete dentition, upper and lower. (2) A head somewhat crushed, and lacking some portions, with full set of upper molars and canines. (3) A great number of fragmentary specimens of different parts of the body.

Dentition.—In the upper jaw the *incisors* are arranged in a semi-circle as in *P. paludosus;* they have long fangs and short, conical crowns, with a decided basal ridge, which is very strong in the third. They increase from first to third, which is very large and pointed.

The *canines* have very large and long fangs (longer than the crowns). The crowns are very stout and pointed, constituting formidable weapons. In section they are nearly circular, with a distinct hinder margin and a slight recurve. A rudimentary anterior margin is sometimes present, and of the faces thus marked off, the inner is much smaller and flatter. The general direction is forwards, downwards, and slightly outwards. There is a diastema of about half an inch between the incisors and canines.

Premolars.—The first is very small, about half an inch from the canine, simple and conical, with an obscurely marked basal ridge, and two heavy fangs.

Further description of the upper dentition is unnecessary, owing to the complete work of Dr. Leidy upon the subject.

The *lower incisors* are placed in a semi-circular row, and are somewhat procumbent, though hardly as much so as in the tapir. They are much as in *P. paludosus*, but are relatively smaller, and have not such distinct basal ridges; features which indicate a departure from the carnivorous type, and a nearer approach to the herbivorous type. The lateral incisor, though the largest of the series, is not so large nor so pointed as in *P. paludosus*. There is no diastema.

The *canines* are about equal in size to those of the upper jaw, growing from stout fangs and permanent pulps. The margins of the crowns are more decided, and are smooth, leaving a very narrow and flat inner face. A feeble basal ridge. When the jaw was closed, the lower canines passed inside and in front of the upper.

The *first premolar* stands immediately behind the canine, is longer than that of the upper jaw, consisting of a single pointed lobe implanted by one fang, slightly recurved with a flat inner and convex outer face. Passing in front of the first upper premolar, it leaves a considerable diastema between this and the second premolar.

The *mandible* approximates in form that of the tapir; the lower border is less curved fore and aft, the alveolar border is slightly concave antero-posteriorly. The molars converge in front, presenting a very different arrangement from that in *Sus*. The ramus is heavy, contracting in depth forward, and very slightly in section. Below the alveolus, on the interior side, the ramus is greatly swollen for two thirds of its depth, to accommodate the very long and strong molar fangs.

The rami converge to the second premolar, where they expand laterally to the canine alveolus. The symphysis is shorter relatively than in *P. paludosus*, and the chin is regularly rounded in front. The mental foramen is below the second premolar. The dental foramen is small, very high up, and far behind the molars. The alveolar border ascends rapidly behind the last molar, expanding laterally into two ridges.

The coronoid is slender and recurved. The condyle is broad, flat behind, inclines forward internally; there is scarcely any depression between the condyle and the coronoid. The masseteric fossa is very wide and deep. The angle of the ramus is broken, but evidently has no such posterior extension as in *P. paludosus*. It thins out rapidly behind.

Measurements.

UPPER JAW.	P. pal. M.	P. major. M.
Length of entire molar series	·147	·170
Length of three true molars	·085	·102
Length of three premolars	·058	·068
Diameter of last molar, transverse	·039	·047
Diameter of last molar, antero-posterior	·036	·036
Diameter of second molar, transverse	·036	·040
Diameter of second molar, antero-posterior	·033	·036
Length of incisor series	....	·075
Fore-and-aft diameter of canine at base	·016	·021
Length of crown of canine	....	·033
Diastema between canines and incisors	....	·013
Diastema between canines and molars	....	·010

LOWER JAW.	P. pal. M.	P. major. M.
Incisor series...	....	·073
Median incisor fang, length........................	....	·034
Median incisor crown, length......................	....	·012
Diameter of median incisor crown, transverse....	....	·009
Diameter of lateral incisor crown, transverse.....	....	·011
Diameter of lateral incisor crown, antero-posterior.	....	·012
Length of fang of canine............................	·055	·052
Extreme breadth of fang of canine.................	·023	·024
Crown of canine, length............................	....	·034
Crown of canine, fore-and-aft diameter at base...	·019	·021
Diameter of first premolar, antero-posterior......	·015	·010
Diameter of first premolar, transverse............	·009	·007
Diastema between first and second premolars.....	....	·014
Length of entire molar series......................	·164	·186
Length of molar series, omitting first premolar...	·132	·163
Diameter of last molar, antero-posterior..........	·017	·050
Diameter of last molar, transverse................	....	·027
Depth of jaw below last molar.....................	·068	·075

Among other specimens obtained at the divide between Henry's Fork and Cottonwood Creek, was a nearly complete but somewhat shattered skeleton of a *Palæosyops major*, found contiguous to the head of the same, parts of which are just described. Portions of these are figured on a one fourth scale in Plate II. The description is from a nearly perfect atlas and axis, several cervical, dorsal, and lumbar vertebræ, more or less complete, and a portion of the sacrum and pelvis, in addition to several bones of the limbs.

The *atlas* has a broad inferior arch, contracted antero-posteriorly, and deeply notched for the prominent odontoid of the axis. The heavy superior arch, inclosing a large and depressed neural canal, is capped by a low tuberosity. It slopes into a flat, and very broad transverse process, which thickens backwards, and is perforated by the vertebrateral canal. Forwards the transverse process is notched for the exit of the first spinal nerve. The anterior or condylar faces are continuous, deeply concave from above downwards; and slightly so from side to side; while the posterior faces are oval, nearly flat, and directed backwards and inwards.

The *axis* is proportionately small, with a long opisthocœlous centrum, depressed, and produced forwards into a stout conical odontoid process, and marked below by a strong hypapophysial keel, which, developed on the latter half, thickens backwards. The wide and thin pedicles inclose a high neural canal, support the posterior zygapophyses and a prominent and peculiar neural spine. The latter is flattened, broad, and recurved posteriorly, thinning rapidly forwards into a prow-shaped recurved edge. The transverse processes arising from the latter half of the centrum are undersized and widely perforated at base. The anterior faces are very broad, directed outwards, and slightly rounded from above downwards; they expand as they diverge (see Fig. 2). The post-zygapophyses are small convex faces, projecting at the base of the neural spine.

The *remaining cervicals*, five in number, are short, all carinate, except the seventh, and opisthocœlous; with the faces expanding only slightly beyond the body of the centrum. A peculiar feature is a small pit upon the convex anterior face, indicating either a ligamentous attachment with the antecedent vertebra, or a remnant of the notochord. This is a feature we have not noticed elsewhere. A long and heavy transverse process supports a large and widely-perforated inferior lamella. The pedicles are quite wide at base, inclosing a large neural canal. The entire upper part of the arch is unfortunately wanting.

The *dorsal* centra are smaller than the cervical, slightly opisthocœlous, and carinate. They are sub-cylindrical anteriorly, with decided costal surfaces before and behind; approaching the lumbar region they become sub-triangular. A high neural arch supports a stout, but never very high neural spine; this projects backwards, keeled in front, expanding and deeply grooved behind. The zygapophyses are small and nearly vertical. There is a short and thick transverse process.

The *lumbar* vertebræ are long, decidedly opisthocœlous, becoming wider and more depressed as they approach the sacrum. The first *sacral* vertebra presents the same characteristics as the last lumbar, only the body of the centrum is slightly shorter. It has a very broad pleuropophysial plate. The remaining sacral vertebræ are broad, and very greatly depressed, rapidly decreasing in size. The transverse processes are slender. The very low neural spines anchylose into a long ridge. The number cannot be ascertained, owing to the fragmentary state of the sacrum. The caudals indicate a tail of not very great length; as the neural canal is small and persists in only a few of the anterior vertebræ.

Measurements of Vertebræ.

	Pal. major. M.
Atlas, width, including transverse processes	·195
Atlas, length of inferior arch	·082
Atlas, height, including superior and inferior arches	·078
Axis, width, anterior articular faces	·110
Axis, length, excluding odontoid process	·058
Axis, height of neural spine	·096
Axis, length of odontoid process	·025
Seventh cervical, length of centrum	·037
Seventh cervical, width of posterior face	·043
Dorsals, anterior region, length	·038
Dorsals, anterior region, width, articular face	·034
Lumbar, width of posterior face	·058
Lumbar, length of centrum	·056
First sacral, width of anterior face	·059

The ribs, of which great quantities of fragments remain, were slender and not of very great width.

The *femur*, which is figured in Plate I., has a small head, supported by a short neck. The shaft, very broad below the head, supports the third trochanter one third of the way down; below which it takes a cylindrical form, and expands slightly above the trochlea. The two condyles, separated by a wide and deep popliteal groove, are nearly sub-equal in size—the internal somewhat the larger, while the trochlea is long, narrow, and symmetrical. The great trochanter rises slightly above the head, and overhangs a long and quite deep digital fossa. The second trochanter is small. The popliteal space is slightly concave.

Tibia.—The proximal end of the tibia is very large, with massive rugosities for muscular attachment. The two proximal faces are sub-equal, slightly convex, and separated by a prominent spine, which is grooved at the top. The shaft is long, straight, and compressed antero-posteriorly; the anterior ridge on upper third is very high, and expands into the tuberosity, which is enormous; while on the posterior face is a deep fossa just below the head. The distal end of the shaft is remarkably small, with two concave faces—the internal the smallest—with a low oblique ridge dividing them. There is a prominent malleolus.

The fibula (which is a distinct bone) has a small proximal end, and expands slightly below, with a large external malleolus.

Measurements of Limbs.

	M.
Femur, total length	·420
Femur, transverse diameter at distal end	·098
Tibia, length	·310
Tibia, transverse diameter, proximal articular face	·090
Tibia, transverse diameter, distal articular face	·077

The *pelvis* is described from the right and left ilia, which are broken as they expand towards the crest; the acetabulum is fragmentary, but all the parts remain. The ischium and pubis are only represented by fragments. As figured in Plate V., the acetabulum is restored from a somewhat smaller specimen. The most striking feature of the *ilium* is the long and somewhat constricted neck. The acetabular border is long and curved forward, thinning gradually as it approaches the crest; and marked on the iliac surface, near the acetabulum, by a slight rugosity for the rectus muscle.

The ischial border, while less arched, is probably longer; and the sacral surface, distinctly defined, indicates that the rounded upper border of the ilium rarely reached above the sacral spines. There is a deep groove between the ischial and pubic borders, the latter disappearing about half way up the iliac surface. The gluteal surface near the acetabulum is much depressed; above it expands into a broad, flat, thin plate.

The *ischium* has a stout neck and thick expansion below; the upper border is very heavy. From what remains of the *ischium* and *pubis* we can infer a large elliptical obturator foramen, a short pubic symphysis, a narrow and rather slight pubis, with a small nearly cylindrical neck. The acetabulum is deep, with a thick well-raised border, and a large, oblong, and very deep pit for the ligamentum teres.

Measurements of Pelvis.

	M.
Ilium, transverse diameter above acetabulum	·062
Ilium, long diameter (estimated)	·232
Acetabulum, greatest diameter	·050
Ischium, transverse diameter below acetabulum	·041
Pubis, diameter at obturator foramen	·035

The *tarsus* is arranged in the usual Perissodactyle order. The *calcaneum* is stout, of good length, tuberous, but not expanding much at the extremity; presenting two faces for the astragalus, of which the internal is the largest, and a large distal face for the cuboid. The *astragalus* is wide, with less asymmetry in the two articular facets than is common. It articulates with both the cuboid and navicular.

Measurements of Tarsus.

	M.
Astragalus, total width, tibial facets	·048
Astragalus, total length	·056
Astragalus, length, navicular facet	·046
Astragalus, width, navicular facet	·035
Calcaneum, total length	·120
Calcaneum, total breadth	·057
Calcaneum, length of tuber calcis	·050

All the above were found at Henry's Fork Divide. In another locality, in connection with a fine head of *P. major*, was found the lower half of a *humerus*.

The *shaft* is cylindrical and twisted, expanding widely at the distal end. Of the two condyles the external is the largest; and, to quote from Dr. Leidy, "a deep supracondylar fossa occupies the front of the humerus, opposed by a deeper and more capacious anconeal fossa." The greatest breadth between the supracondyloid eminences is ·091 m.

PALÆOSYOPS PALUDOSUS, Leidy.

Cont. to Ext. Vert. Faun, *P. lævidens* Cope. U. S. Geol. Survey of Terrs, 1872, p. 591.

Specific characters.—Second superior molar has but one outer tubercle. The cones are low as compared with *P. ma-*

jor. The cingula are much less developed. The angle of the lower ramus is much longer. The zygoma is massive and wide. An internal process on the glenoid cavity. Temporal fossæ very deep. Crest of occiput nearly in same vertical line as occipital condyles.

The full details of *P. paludosus*, which follow, may seem somewhat unnecessary, to one familiar with the comprehensive work of Dr. Leidy upon the subject. Care has been taken not to retrace any of his steps; and the descriptions given below are of those parts of the animal which he did not possess at the time of writing. They include (1) a cranium perfect posterior to the orbits; (2) a right *ramus mandibuli*, with full dentition, except the first premolar; (3) portions of the fore-limbs, scapula, and pelvis, pes and manus, and many fragments of other parts.

Among the first discoveries on Cottonwood Creek, in a stratum of fine green sand, was a head of *P. paludosus*, complete posteriorly, and broken off just behind the orbit. A rear view of this has been admirably figured in Plate I.

General appearance.—The base of the cranium is of great width; the occiput is high, inclined very slightly backwards, and deeply concave from side to side. The forehead is triangular and narrow. The temporal fossa is of immense size and depth, leaving a small intracranial cavity. The zygomas are heavy, and arching widely outwards give this part of the head a strong resemblance to the *Felidæ*.

Description in detail.—The narrow but prominent *basi-occipital* segment is broadest posteriorly and tapers forwards; divided by a median ridge, which expands anteriorly into a large tuberosity; a feature also characteristic of the tapir. About half an inch in advance of the condyles are the condylar foramina. The *condyles* are formed of the ex-occipitals, which are low and of great lateral expansion. Their junction with the mastoids is marked by a large foramen. The *par-occipitals* are short and styliform. The *supra-occipital* region is very large, high, and deeply concave from side to side, much wider than in the tapir, with a marked interparietal suture. The *condyles* are wide, but not very deep, approaching each other very closely below. The *basisphenoids* are long and narrow, tapering forwards. The *alisphenoids* are mutilated, but indicate large vertical ridges joining the parietals, and heavy pterygoid processes, perforated at the base by the alisphenoid canal. They are again perforated by the foramen ovale three fourths of an inch behind this. This completes the base of the skull.

The *parietals* are very large, they form nearly the whole of the temporal fossæ; which, deep and wide, enclosing a small cranial cavity, contrast strongly with the long and shallow temporal fossæ of the tapir. The parietal crest is very broad, and grooved at the top. This high crest and deep adjacent temporal fossæ we at first mistook as pointing to an exclusively carnivorous type. The forehead has a triangular appearance, from the divergence of the two side ridges of the sagittal crest. The *postorbital* processes are very large, but do not reach the opposing processes of the *malar*. The orbit is thus left incomplete posteriorly, while in form it greatly resembles that of the *Sus*. The *squamosal* encroaches considerably upon the temporal; and sends outwards and downwards a great zygomatic process, which arches outwards from the skull as in the *Felidæ*, and is more powerful than in any living carnivore (Leidy.) A strong downward direction is especially characteristic, the whole describing a sigmoid curve. The *mastoids* are of great size vertically, and transversely they are confluent with the paroccipitals. A low, thick process on the internal side of the glenoid cavity prevents lateral motion.

The *nasals* are long, broad, and thick, convex from side to side, narrowing slightly anteriorly. They are straight, as in *Sus*, which they resemble more than they do either tapir or rhinoceros. The anterior borders are rounded, and do not reach as far forward as the symphysis of the premaxillary. The *malar* is broad and thick, probably forming but little of the face, being directed downwards and backwards to meet the zygoma. The postorbital process is short, and rather larger than Dr. Leidy has indicated. The *maxillaries*, smaller proportionately than in *Sus*, form posteriorly the floor of the orbit; while the infraorbital foramen is situated over the last premolar. The premaxillaries fail to reach the nasals.

Comparative Measurements of Head.

	P. paludosus.	P. major.	Tapir.
	M.	M.	M.
Height of occiput	·122	·148	·120
Breadth of occiput at post tympanic processes	·160	·166	·110
Breadth of cranium at ends of post-glenoid processes	·172	·210	·126
Transverse diameter of occipital foramen	·031	·049	·040
Vertical diameter of occipital foramen	·027	·033	·027
Depth of occiptal condyles	·033	·039	·029
Breadth of occiptal condyles	·040	·049	·041
Breadth at occipital condyles together	·082	·100	·082
Width of basi-occipital at anterior condyloid foramina	·038	·039	·028
Width of basi-occipital at junction with basi-sphenoid	·025	·032	·020
Width of crest dividing the temporal fossæ posteriorly	·011	·020	·011
Breadth of cranium outside of zygomata	·262	·280	·180
Depth of zygoma	·046	....	·036

Lower jaw of *P. paludosus* described from another specimen.

The peculiar feature of this jaw is its remarkable extension back of the molar series (a feature which has not been noticed heretofore), the distance from the last tooth to the angle being greater than the length of the entire molar series. From beneath the last molar, the lower margin curves gently up to the symphysis. The alveolar border is but slightly curved. Behind the last molar the lower margin forms a sigmoid curve, much more decided than in *P. major* (first upwards and then downwards), and the ramus thins out rapidly to the angle, where it has a slightly raised border; at the diastema the ramus curves outwards so as to throw the canines out of the line of the molars. The symphysis is long.

Dentition.—The *incisors*, three in number, from length and shape indicate a semicircular arrangement, as in most Perissodactyles. They increase from first to third. The crown of the first resembles very much that of the ruminants, but has a straight posterior basal ridge. Its forward edge is worn so as to expose a small tract of dentine. The second is larger, but very much like the first in conformation; while the third has an acute conical crown with a strong basal ridge.

Of the *premolars* the first is wanting in this specimen. The second is bilobed, the anterior lobe much the larger. It has a small accessory tubercle on its anterior slope. The outer face is rounded, the inner nearly flat. The posterior lobe is very small and obtuse, and with the inner face slightly grooved. In third and fourth the valley between the lobes and the groove on their faces enlarge; while the anterior lobe becomes relatively smaller.

Measurements of Lower Jaw.

	M.
Diastema between canine and first premolar	·024
Length of entire molar series	·143
Distance between last molar and extremity of angle	·147
Depth of jaw at last molar	·070

Among the first discoveries on Cottonwood Creek, in connection with the lower jaw of *P. paludosus* just described, were an ulna and radius, and part of the manus and scapula of the same. These have been figured, together with other fragments belonging to the same species, in Plate III., and form an interesting complement to the nearly complete hind limb of *P. major* previously described.

The *scapula* has a shallow oval glenoid cavity, which is concave longitudinally. Separated from it by a narrow notch is a strongly recurved coracoid, placed on the inner side of the bone. Above the coracoid the border is thin, and arches inwards and then forwards; while the glenoid border is much straighter, rising from a slight tuberosity. The spine rises gradually from the glenoid, and does not bear any indication of an acromion.

The *humerus* is represented by proximal and distal extremities. The latter presents the same characters on a smaller scale as in *P. major*. Upon the proximal end there is a deep bicipital groove. The greater tuberosity has a hooked anterior projection, and runs back into a ridge behind. (See Fig 7, Plate III.)

The proximal end of the *ulna* (of median breadth anteriorly) narrows into a prominent ridge behind. This gives a trihedral character to the shaft. This is persistent, but is less marked as the fore-and-aft diameter decreases. The remaining features of note are a high tuberous olecranon expanding behind; a shallow sigmoid, divided by a deep fossa into two long divergent articular faces; a trihedral shaft tapering slightly, but distinct from the radius; a narrow distal extremity, with a small irregular face for the cuneiform.

The *radius* has a rather short shaft, curved forwards, expanding below, and placed immediately in front of the ulna. A strong ridge runs from the external tuberosity up the posterior face of the shaft, disappearing two inches from the top. The proximal end, fitting against the ulna by two small faces, forms the greater part of the elbow-joint, a strong median ridge dividing the proximal face into two subequal surfaces with well-raised borders. The distal end of the shaft is rugose, and more prominent than the proximal. It has an oblong transverse face, which is slightly concave fore and aft. The position of the radius, immediately in front of the ulna, and its manner of articulation, remove any possibility of rotation, a feature in strong contrast with what we should expect from the character of the head.

Measurements of Ulna and Radius.

	M.
Ulna, length, excluding olecranon	·224
Ulna, length of olecranon	·080
Ulna, transverse diameter, proximal surfaces	·041
Ulna, diameter of shaft at median line	·038
Ulna, fore-and-aft diameter, distal face	·025
Ulna, transverse diameter, distal face	·017
Radius, proximal face, transverse	·048
Radius, proximal face, fore and aft	·030
Radius, median diameter, shaft	·019
Radius, median diameter distal face, transverse	·043
Radius, median diameter distal face, fore and aft	·016

The *manus* is described from the third and fourth metatarsals, with three phalanges. The metatarsals are short, with a wide, stout shaft, flat in front and slightly arched forwards behind. The proximal ends unite and form a continuous articular face, concave from side to side; while upon the exposed sides are smooth facets, indicating the presence of two additional toes of less size. (See Plate III.) The convex distal faces are marked behind by a slight groove.

The *phalanges* are very short and broad, the lower facets marked by a shallow median groove. The ungual phalanx is wide and short, with a crescentic outline.

Measurements.

	M.
Third metatarsal, total length	·090
Third metatarsal, transverse diameter, proximal face	·017
Third metatarsal, transverse diameter, distal end	·019
Second phalanx, length	·024
Third phalanx	·014
Total estimated length of manus (6 inches)	·155

PALÆOSYOPS VALLIDENS, Cope.

Proceedings Am. Phil. Society, 1872, p. 487; Hayden's Survey, 1872, p. 572.

Specific characters.—Founded on details of dental structure. Superior size. In the upper molars two strong transverse ridges connect the inner tubercle with the outer crescents, inclosing a pit between them. In the premolars the outer crescents fuse almost into a single ridge. These united crescents are relatively narrower. The inner molars are also narrower, and the posterior tubercle of the last is an elevated cone.

Fragment of a lower jaw containing the two posterior molars. The measurements of the teeth are identical with those given by Professor Cope. The depth of the ramus below the last molar is three and one third inches, and it is slightly heavier below than in *P. major*.

Restoration of Palæosyops.—The elevation of *Palæosyops major*, which is taken as a type, was approximately the same as that of *Palæotherium magnum* as restored by Cuvier.

The nearly complete ilium, femur, tibia, and tarsus of *P. major* in our possession, enable us to make the following estimate of the elevation of the hind quarters.

	INCHES.
Pelvis, height of ilium above acetabulum..........................	9
Femur and *tibia* by actual measurement, allowing for bend at the knee.	26
Tarsus, from astragalus, and cuboid....	3
Metatarsus and *phalanges* (estimated from manus of *P. paludosus*.....	8
	46

It is probable that the fore-shoulders were of the same height. From the atlas, axis, and from other cervicals, several dorso-lumbar vertebræ, and part of the sacrum, a rude estimate places the length of the animal at sixty-seven inches, or nearly six feet, including the head, and excluding the tail.

In general features it strongly resembled the tapir, with stout body, slender tail, and very short neck, compensated by a proboscis of considerable length. In comparing the heads of the ancient and modern representatives of this class of Perissodactyles, the points of contrast are the wide stout zygomas, the deep temporal fossæ, the protruding nasals, and the narrow gaps in the dental series of the older type.

There are also strong points of resemblance in the structure of the *Palæosyops* and *Palæothere*. Some of the cranial homologues have been detailed by Dr. Leidy. The femur of the Palæothere is shorter, the tibia longer; they are both much stouter. The tarsus of the *Palæosyops* is narrower, while the remainder of the Pes is very similar. The pelvis, and particularly the ilium and acetabulum of *Palæosyops*, is more palæotheroid than tapiroid. The similarity of the fore-shoulder in the two types is also striking. The neck of the *Palæosyops* was probably shorter.

In the descending series of *Palæotheres* terminating in the diminutive *P. minus*, and characteristic of the upper eocene of France, we have an interesting counterpart in the large family of *Palæosyops*, of equal diversity of size and characteristic of our lower eocene, and it is hoped that future discoveries will render a complete discussion of this interesting coincidence possible.

LIMNOHYUS, Leidy.

Proceedings Academy Natural Sciences, Phila., 1872, p. 242; Palæosyops; Marsh, American Journal Sci. and Arts, 1872, p. 122; Hayden's Survey, 1872, p. 592, Cope.

Generic characters.—Resembles *Palæosyops* in general features. Distinction founded on the possession of two conic tubercles, of the inner series, on the last superior molar instead of one.

LIMNOHYUS LATICEPS, Marsh.
Palæosyops Laticeps. Am. Journ. of Sc. v. iv. p. 122.

Specific characters.—" Cranium is broad. The zygomatic arches much expanded. The nasals are narrow and elongated, and more like the corresponding bones in Hyrax than those in the larger pachyderms."

Fragments of the upper molar and premolar series. The species is determined by measurements given by Dr. Leidy of the second upper molar.

LEUROCEPHALUS, *Gen. Nov.*

Established on a specimen having a nearly complete dentition, and portions of the cranium.

Dental Formula: (I. $\frac{3}{3}$, C. $\frac{1}{1}$, Pm. $\frac{4}{4}$, M. $\frac{3}{3}$,) × 2 = 44.

Upper incisors acute, with strong posterior ridges, lower incisors compressed and laniariform, canines compressed, with serrated cutting edges; first upper premolar with rudimentary anterior lobe, last upper molar with rudimentary posterointernal cusp. Molars constructed as in *Palæosyops*, but higher, with sharper cones and more erect external lobes. Internal median valley very much deeper. Little or no depression at the forehead; zygomatic arch round, comparatively straight and does not project outwards, and with obscure postorbital process. Premaxillaries short and straight. Mandible with nearly straight lower margin, and shallow masseteric fossa; mental foramen single.

LEUROCEPHALUS CULTRIDENS, *Sp. Nov.*

This species was considerably larger than *Palæosyops major*, with which it shows affinities, as also with *Titanotherium Proutii*. See Plate IV.

The *Frontal* is narrow and flat, or slightly arched. It sends out a strong curved postorbital process, which projects outwards, but not so much outwards as in *Palæosyops*. The sagittal crest runs back from the postorbital in a low rounded ridge, rising but little as it recedes. The superciliary ridge is distinct, and is pierced by two small venous foramina. The shape of the entire bone is much as we find it in the dog, except that it does not arch so much. One of the most marked differences between this genus and *Palæosyops* is here shown. In the latter the forehead rises abruptly from behind the orbit, while in the former there is only a slight rise for some distance behind the orbit. As the bone is broken here, it is impossible to state whether there is any rise at all. Probably not, however. Below the post-orbital process, on the lateral aspect of the bone, there is a low ridge which separates the orbit from the temporal fossa. The under surface of the frontal is smooth.

The *Nasal*, of which but a small portion is preserved, is flat on top, but bends downward at the angle.

The *Maxillary* is long and stout, over the last and penultimate molars it broadens to form the floor of the orbit, which is of unusual size. Between the canine and the malar the maxillary arches inwards, forming a perceptible concavity, at this point it reaches its greatest vertical height, as it rises to join the nasal. The alveolar border is curved in two directions, one with the convexity outwards, and the other downwards. It exhibits no emargination in the diastema between the canines and molars. The palatine plate is long, thick, and narrow; it is flat fore and aft, but concave transversely, owing to the elevation of the alveolus; the suture also is raised slightly. In thickness it varies; being thickest between the canine and the second premolar, and beyond this becoming thinner. Its forward termination seems to be obtuse, running for a short distance along the premaxillaries. The infraorbital foramen is large, situated over the fourth premolar, lower down, and nearer to the malar than in *Palæosyops*.

The *Premaxillary* is rather short and stout; it is slightly compressed, and ends in a sharp keel on top. Although not so thick as in *Palæosyops*, it is much larger vertically. The inner surface is ridged; it has no palatine process, and no spine, so that the incisive foramen is large and undivided. There was no symphysis between the two premaxillaries, they do not show even any articular faces for each other, so that they may not have been in contact during life. The animal was adult, but not old. The incisive alveolus is short and straight, and is but little out of the line of the molars. There is a very long diastema between the incisors and the canine, and here the premaxillary is deeply notched to allow the passage of the lower canine.

The *Malar* is very different in shape and size from the same bone in *Palæosyops*. Its facial extent is greater, as it articulates with the maxillary as far forward as the *first* molar. It is not so broad from above downwards, but is considerably thicker; the articulation with the squamosal is by flat surfaces. The postorbital process is small and indistinct. The orbit is of great size; its breadth, shown by the maxillary floor is unusual, but its fore-and-aft diameter is extreme, being nearly twice as great as in a large *Palæosyops major*.

The *Squamosal* is a short, slender, and trihedral bone; it is not at all like the squamosal of *Palæosyops*. It is smaller even than that of *P. paludosus;* its outer margin is not turned up, making the upper surface concave; it does not project much outward, and scarcely at all downwards, as this is rendered unnecessary by the flatness of the forehead. The chief difference, in the size of this bone, between this genus and *Palæosyops* is found in the much greater vertical and lateral diameter of the latter; in length they are about equal. The glenoid cavity is large, shallow, and transverse.

As a whole, the zygomatic arch is comparatively slender; it is rounded, and does not project much laterally from the side of the head; but its chief peculiarity consists in the fact that it is nearly horizontal. It is thus altogether different from the arch of *Palæosyops*.

The *Occipital Condyles* are different from those of that genus. They are flatter, shorter from above downwards, and proportionately much smaller; but are expanded laterally in much the same manner. At the border of the foramen magnum they are gently rounded instead of having a sharp angle. These borders are segments of circles, but as the condyles are broken from the rest of the occiput, we cannot infer from this the shape of the foramen.

The *Mandible* is long, stout, but comparatively shallow. It is of nearly uniform thickness throughout, and does not exhibit the thickening of the alveolus and thinning, of the lower margin, which is shown in *Palæosyops*, especially in *P.*

major; and the downward and inward arching of the alveolar border is not marked to the same extent as in that species. The lower border is nearly straight, and has but one slight emargination just behind the symphysis. The ascending portion of the ramus rises near the last molar quite abruptly, and makes an angle of nearly 90° with the horizontal portion. The masseteric fossa is shallow and obscure, and the anterior bounding ridge is almost obsolete. This part of the jaw is very different from any of the species of *Palæosyops*. The symphysis is quite long, and extends back as far as the second premolar; the chin is probably rounded. The dental foramen is situated high up in the ascending portion of the ramus; the mental foramen is placed below the second premolar; it is large and single.

Cranial Measurements.

	M.
Breadth of frontal at postorbital process	·045
Breadth of frontal in front of postorbital	·030
Breadth of nasal just in front of orbit	·032
Length from angle of nares to end of premaxillary	·1235
Length of premaxillary	·056
Length of maxillary	·230
Breadth of palatine process at canine	·026
Breadth of palatine process at second premolar	·029
Vertical height of maxillary at angle of nares	·064
Breadth of maxillary across floor of orbit	·058
Fore-and-aft diameter of orbit	·054
Length of malar along maxillary	·094
Vertical diameter of malar behind maxillary	·029
Transverse diameter of malar behind maxillary	·033
Fore-and-aft diameter of glenoid cavity (about)	·029
Vertical diameter of squamosal at end of malar articulation	·032
Transverse diameter of squamosal at end of malar articulation	·0325
Length of ramus mandibuli from symphysis to ascending portion	·177
Depth of jaw at last molar	·068
Depth of jaw at second premolar	·061

DENTITION.

UPPER JAW.—The *incisors*, three in number, are placed close together in a nearly straight line, which does not make so great an angle with the line of molars as in *Palæosyops*. They increase regularly in size from the first to the third, which is very large. They all have stout rounded fangs, and sharp pyramidal crowns, with strong basal ridges separated from the acute blades by deep clefts. Between the incisors and the canine there is a long diastema.

The *canine* is of great size; it has a swollen fang, and a long recurved crown which is much compressed, and has sharp serrated cutting-edges. The external face is the more convex, the internal has a well-marked basal ridge, and both are longitudinally striated. The canine is separated from the premolars by a short diastema.

Premolars.—The first premolar stands by itself, separated by a short interval from the second. It is implanted by two fangs, and has a compressed crown with sharp cutting-edges. It is nearly twice the size of the corresponding tooth in *Palæosyops*, which is simple and conical. In the specimen under description there is, besides the principal lobe of the tooth, a rudimentary anterior lobe which gives the crown an elongate shape. The basal ridge is shown on the inner side only. The second premolar has a crown which is oval in section. The external part consists of two sharp-pointed cusps, separated by a valley, but confluent at base; while the internal division is a low ridge (not a pointed cone as in *Palæosyops*) with a tubercle developed behind it. The basal ridge is marked all around, and sends up a buttress along the antero-external lobe; as is also the case in the third and fourth premolars. These are enlarged copies of the second; but have the internal lobe elevated into a sharp cone, and showing a much stronger basal ridge and deeper median valley. These teeth differ in several respects from the premolars of *Palæosyops*. The crowns are higher and the lobes sharper; the basal ridge is more distinct, and is not interrupted at the internal cusp; the external ascending buttress is stronger; and there is no fold between the external lobes of the fourth.

The *molars* are constructed in a manner which resembles that of *P. paludosus* rather than of *P. major*, but its resemblance to the form shown in *Titanotherium* is stronger than to either. They have broad, square crowns, which increase progressively in size from first to last. There is a convexity running up the median line of the external lobes, and the two posterior lobes are connected together at their bases. The first molar does not exhibit such a decided increase in size over the last premolar as is seen in *Palæosyops*; the two external lobes are more erect and sharper than in any form of that genus. The cingulum is complete even across the internal cones, and the median valley is very deep, as in *Titanotherium*. There is a conspicuous fold at the junction of the external lobes, and a minute tubercle between the two anterior lobes; this tubercle is present only in the first molar, and not throughout the series, as in *Palæosyops*. The second molar in general form is like the first; but is larger, has a deeper median valley, more conspicuous external folds, and a stronger basal ridge. There is also some change in the proportions of the crown; the first measuring the same in both directions, while the second is wider than it is long. The last molar has but one internal cone; the second is represented by a tubercle; which however is a true rudimentary lobe and not a development of the basal ridge. In this tooth, the external fold is very large and the median valley very deep.

LOWER JAW.—The lower molars and premolars are very much like those of *Palæosyops*, but are longer, narrower, and higher. The *incisors* are not at all like *Palæosyops*, but are compressed and laniariform, with acute crowns and sharp cutting-edges. The canine is not present in our specimen; but that it was large, is shown by the long diastema between the upper canine and incisors.

Premolars.—The first is small, simple, and implanted by a single fang close to the canine. The diastema between it and the second premolar is about one half of that in *P. major*. The second premolar is the most peculiar tooth in the lower series; as it carries the development of the anterior lobe at the expense of the posterior lobe, to a still greater extent than in *Palæosyops*, forming a crown like the blade of the carnivorous sectorial. The third lobe of the last molar is no longer a simple cone, but a functional lobe, with two crests running from it, which bound a median valley.

Teeth Measurements.

UPPER JAW.	M.
Length of entire dental series	·275
Length of incisor series	·0375

Length of diastema between canine and incisors	·018
Fore-and-aft diameter of canine	·019
Transverse diameter of canine	·015
Length of diastema between canine and premolars	·0145
Length of premolar series	·083
Length of first premolar	·016
Width of first premolar	·009
Length of second premolar	·021
Width of second premolar	·017
Length of third premolar	·022
Width of third premolar	·022
Length of fourth premolar	·024
Width of fourth premolar	·027
Length of true molar series	·106
Length of first molar	·031
Width of first molar	·031
Length of second molar	·037
Width of second molar	·040
Length of third molar	·038
Width of third molar	·041

LOWER JAW. M.

Length of entire molar series	·202
Length of premolar series	·081
Length of true molar series	·121
Length of second premolar	·024
Width of second premolar	·0125
Length of third premolar	·022
Width of third premolar	·014
Length of fourth premolar	·0235
Width of fourth premolar	·0155
Length of first molar	·030
Width of first molar	·017
Length of second molar	·038
Width of second molar	·021
Length of third molar	·053
Width of third molar	·024

Found at Henry's Fork Divide.

This may eventually prove to be a species of *Telmatherium*, (Marsh); but the description given by him of that genus is so brief and uncharacteristic that it might apply to any of the allied genera. Indeed, Dr. Leidy has regarded it as a synonym of *Palæosyops*.

HYRACHYUS, Leidy.

Proc. Ac. Nat. Sc., 1871, p. 229, *et sqq.*

A genus of tapiroid mammals, which is thus defined: "*Lower Jaw:* Third molar with two crests; four premolars, third and fourth with one transverse and one longitudinal crest. *Upper Jaw:* Seven molars, first without interior heel; premolars with two transverse crests." (Cope.)

Synopsis of Species.

I. Long diastema between lower canines and molars.
   A. Lower jaw with six molars................*H. nanus.*
   B. Lower jaw with seven molars.
      a a. Upper molars with complete cingulum. Enamel wrinkled.
        a. The last molar with two external lobes on nearly the same transverse line........*H. intermedius.*
        b. Last molar with the external lobes on the same longitudinal line................*H. modestus.*
      b b. Upper molars with cingulum incomplete externally.
        a. With anterior conical buttress larger than anterior lobe.
            α. Buttress separate, and no fold from anterior lobe............*H. princeps.*
            β. Buttress united, and quite large fold from anterior lobe.....*H. imperialis.*
        b. With anterior conical buttress smaller than anterior lobe.
            α. Tooth large...............*H. eximius.*
            β. Tooth smaller............*H. agrarius.*
      c c. Cingulum only on outer side of first molar, buttress separated by a ridge from anterior lobe, fold from this lobe very prominent.
            H. implicatus.

II. Short diastema between lower canines and molars.
   A. Descending ridge from antero-external point of lower molar extends entirely across the crown of the tooth anteriorly....................*H. crassidens.*

HYRACHYUS IMPERIALIS, sp. nov.

Established on the second and third molars of each upper jaw, and three premolars, and one lower molar, portions of the skull and vertebræ.

This species is, with the exception of *H. princeps*, (Marsh), the largest species of the genus as yet known. It differs from *H. princeps* in the greater proportionate length of the molars, in their larger size, compared with the bulk of the animal, and in the less separation of the anterior external conical buttress. The third molar is the largest; the fold extending from the antero-external lobe into the valley is very conspicuous. The basal ridge (which is very strongly marked on the posterior edge), is very feebly indicated on the exterior. The transverse crests are more oblique in the second than in the third, but in the latter the valley is deeper and the crests slightly longer. The anterior conical buttress is of great size, being larger than the anterior lobe. The premolars have, as usual, one internal cone; the anterior and posterior external lobes approach close together, while the accessory fold from the anterior lobe becomes very prominent, reaching the internal cone and enclosing a valley between itself and the transverse crest. The basal ridge does not extend around the inner part of the crown as in *H. agrarius*.

The only lower molar we have, seems to be a penultimate molar of the right side. The basal ridge is large anteriorly, very short posteriorly, but does not appear at all upon the sides, as in *H. agrarius*.

The remaining bones of the skeleton indicate a rather small animal, the teeth are therefore proportionately very large.

The occipital condyles are small and sessile; they are much extended laterally, and the external angles are pointed. In other respects they resemble the condyles of *Palæosyops*. The zygomatic arch is slender and does not curve outward strongly. We cannot tell how far the malar encroaches upon the face. After leaving the maxillary it becomes much compressed; it has no post-orbital process. The maxillary is of unusual breadth, extends far backwards, and it forms the floor of the large orbit. The mandible is slender and compressed, with a large dental canal.

The atlas is small, and has narrow transverse processes, which are perforated by a large vertebraterial canal. The dorsals have considerably depressed centra, and the ribs are slender.

Measurements.

	M.
Length of last molar	·025
Breadth of last molar	·030
Length of second molar	·024
Width of second molar	·0305
Length of last (?) premolar	·018
Width of last premolar	·0245
Length of second lower molar	·025
Width of second lower molar	·018

From Henry's Fork Divide.

HYRACHYUS INTERMEDIUS, *sp. nov.*

Established on the upper true molar series of the right side.

This species is intermediate in size between *H. agrarius* and *H. modestus*. As in the latter species, the basal ridge extends around the entire crown of the tooth, with a small interruption on the antero-external lobe. It is more pronounced in front and less behind than in *H. imperialis*. The anterior conical buttress is not nearly so well developed as in *H. imperialis;* the valleys are wide, and the descending fold from the antero-external lobe is obscure, scarcely marked at all in the first. The transverse crests are strongly arched forwards, and the anterior is very much the longer. The peculiar feature of the species is the position of the postero-external lobe, which is pushed to the posterior aspect of the tooth; this is carried so far in the last molar that the two external lobes stand on nearly the same transverse line. The antero-external lobe is much the highest point in the tooth, and in the first molar is remarkably sharp. The enamel is considerably wrinkled, but not nearly so much as in *H. modestus*.

Measurements. M.

Length of first molar	·012
Breadth of first molar	·0155
Length of second molar	·015
Breadth of second molar	·017
Length of third molar	·0145
Breadth of third molar	·017

From Cottonwood Creek.

HYRACHYUS NANUS, Leidy.
Pr. Ac. Nat. Sc., 1872, 20.

Henry's Fork.

HYRACHYUS AGRARIUS, Leidy.
Pr. Ac. Nat. Sc., 1871, 229.

Henry's Fork.

HYRACHYUS EXIMIUS, Leidy.
Pr. Ac. Nat. Sc., 1871, 229.

Cottonwood Creek.

HYRACHYUS CRASSIDENS, *sp. nov*.

Established on the left and a portion of the right ramus mandibuli, containing the first molar.

The jaw presents some differences from the form common in Hyrachyus. The alveolus is more curved, especially near the ascending portion of the ramus; the jaw is narrower, and is especially contracted near the incisive alveolus. The lower margin of the ramus is but slightly curved. The ramus itself thickens regularly from before backwards, and is thickest at the last molar; beyond which it becomes thinner again. The depth of the jaw is comparatively small, and the teeth have a forward inclination; the mental foramina are not so large as in other species of *Hyrachyus*, and are four in number.

The incisors are small, arranged in a semicircle, and have very much compressed fangs, with somewhat procumbent crowns; the canines are small, and form a continuous series with the incisors. There is a diastema between the canine and first premolar; but one which is shorter than in known species of the genus, and which does not indicate the usual curve of the upper border of the ramus.

The molar series, (so far as can be judged from their broken condition), are much the same as in other species; except, perhaps, in the greater relative size of the posterior lobes, and in the more complete closing of the median valley by the junction of the two lobes externally. The ridge which curves downwards and inwards from the external point of the anterior lobe is unusually well developed, and curves across the entire crown. The cingulum is feebly shown; it appears on the anterior and posterior ends of the tooth (first molar) and for a very short distance on the outside of the tooth opposite the median valley; there is no trace of it on the inner side.

Measurements.

	M.
Length of entire molar series (about)	·080
Length of premolar series	·033
Length of true molar series (about)	·047
Fore-and-aft diameter of fang of canine	·008
Distance between canine alveoli	·009
Length of diastema	·115
Length of the first molar	·015
Breadth of the first molar	·010
Depth of ramus below last molar	·0245
Depth of ramus below last premolar	·0215
Greatest thickness of ramus	·017
Length of symphysis (about)	·030

Found at Henry's Fork.

HYRACHYUS ——— ?

There are three premolars of the upper jaw which we cannot refer to any known species; but they are too imperfect to justify the formation of a new species for their reception.

What appears to be a first premolar is inserted by a single broad fang; it is too much worn to show the characteristic enamel folding, and is peculiarly broad and short. The second premolar shows the usual foldings of *Hyrachyus;* save that a distinct valley enters the tooth from the posterior edge, running forward to the transverse crest, and inclosed by the curving backwards of the internal cone. The crown is nearly square; the basal ridge is scarcely marked at all.

Measurements.

	M.
Length of first premolar	·011
Breadth of first premolar	·016
Length of second premolar	·012
Breadth of second premolar	·014

HELALETES, Marsh.

Am. Journ. Sc. and Arts, 1872, p. 218.

Additional material enables us to give further characteristics of this genus, which may be thus defined: Mammals, allied to *Lophiodon*, having a third lobe to the last lower molar, short symphysis, and with the teeth of the lower jaw forming a continuous series. Astragalus has narrow, oblique condyles. Dental formula: I.$\frac{?-?}{3-3}$, C. $\frac{1-1}{1-1}$, Pm., $\frac{4-4}{4-4}$, M. $\frac{3-3}{3-3}$.

Synopsis of Species.

A. Teeth small; a small tubercle on the outer margin, between the cusps of last upper molar..........*H. boöps*.
B. Teeth much larger and proportionately broader; no tubercle on last upper molar..............*H. latidens*.

HELALETES LATIDENS, *sp. nov*.

Lower Jaw.—The ramus is stout, and resembles in general form that of *Hyrachyus*, but the alveolar border is straighter than in that genus; the ascending portion forms a right angle with the horizontal portion, and the masseteric fossa is deep. The symphysis is comparatively short, and there is complete bony union between the rami, which do not diverge so much as in *Hyrachyus*. As in that genus, there are several small mental foramina beneath the premolar series. The dental series forms a nearly continuous line, the only diastema being a very short one between the first and second premolars. The *incisors* have compressed fangs, and are arranged in a slight curve. The canine is small and subcircular at base. The *premolars* also are small; they somewhat resemble those of *Hyrachyus* in shape, but are smaller, and have the posterior lobe not so well developed. The first is implanted by a single fang, the others by two. The *molars* (except the last) consist of two pyramidal lobes, which form transverse cutting-crests, as in *Tapirus* and *Hyrachyus*. The basal ridge is shown on the anterior and internal sides only. Externally the lobes meet at their bases, inclosing the valley on that side. The last molar is by far the longest of the series, and consists of three lobes, of which the anterior is the largest, and the posterior the smallest.

The basal ridge at the anterior part of the tooth is very strongly marked.

Upper Jaw.—The molars increase in size from first to last with considerable regularity. The premolars are small, and, as in *Hyrachyus*, have one single internal cone, which is joined by transverse ridges from the two outer cusps; the anterior accessory tubercle is well developed. The first molar is considerably larger than the last premolar; it is much worn, and does not show anything more than that there are two internal cones. The anterior conical buttress is very largely developed throughout the true molar series. The second and third molars resemble almost precisely the corresponding teeth of *Hyrachyus*. Their shape, however, is more nearly square than is usual in that genus.

Measurements.

UPPER JAW.

	M.
Length of true molar series	·036
Length of last molar	·0135
Width of last molar	·015
Length of second molar	·012
Breadth of second molar	·015
Length of first molar	·0105
Breadth of first molar	·012
Length of last premolar	·009
Breadth of last premolar	·011
Length of third premolar	·008
Breadth of third premolar	·009

LOWER JAW.

Length of true molar series	·042
Length of last molar	·018
Breadth of last molar	·009
Length of second molar	·013
Breadth of second molar	·008
Length of first molar	·012
Breadth of first molar	·008
Length of third premolar	·008
Breadth of third premolar	·005
Diameter of canine at base	·006
Length of symphysis (about)	·027

Found at Henry's Fork.

ARTIODACTYLA.

GENERA INCERTÆ SEDIS.

ITHYGRAMMODON, *gen. nov.*

Our fragments of this fossil show an animal about the size of the llama, and approximating more closely to the *Tylopoda* than to any living forms.

The genus is established upon the two premaxillary bones, containing the incisors, parts of the maxillaries, the canine and the first premolar; besides fragmentary portions of the palatine plates. As the peculiar features of these remains render them unique, we are warranted in establishing the genus *Ithygrammodon*.

Generic characteristics.

The premaxillaries are long, narrow, and very straight, bent slightly in on their own axis as in *Camelidæ*, but expanded laterally at the end of symphysis as in *Pecora*. The upward processes are sharply defined, with a wide, rounded upper border.

The upper incisors are six in number, and *are placed nearly in a straight line fore and aft*, separated from each other and from the canine by small and subequal diastemas.

The *incisors* are nearly equal in size, cylindrical in shape, are placed vertically in their alveoli, and are slightly recurved.

The spine of the premaxillaries is long and thin, giving very long anterior palatine foramina, as in *Ruminantia*. The premaxillaries, in position, are close together, showing a very narrow muzzle.

The maxillaries have two well-developed canines, with long, laterally compressed fangs, recurved and becoming circular in section at the crowns. The diastema between the canine and the first premolar is proportionate to that in *Procamelus*.

ITHYGRAMMODON CAMELOIDES, *sp. nov*.
Specific characteristics.

The *premaxillaries* (See Plate X., Figs. 1-4) are long, with a wide, strongly upward curving process; from the premaxillary symphysis to the third incisor, they increase in thickness; then narrow slightly to the maxillary symphysis. The upper margin of the process is smooth and narrow, curving very gradually upwards; while the lower margin runs more abruptly up, giving to it a strongly pointed curvature.

The outer surfaces are smooth and slightly convex; the inner are marked by a wide deep canal running from just behind the premaxillary symphysis into the maxilla.

The great size of this groove may perhaps be a generic characteristic; as it is much larger than in *Protolabis* (Plate X., Fig. 6), the modern camels, or any of the other ungulates, recent or fossil, that have come under our notice.

The premaxillaries do not co-ossify, as the spines are smooth on their inner side. Just outside the spines, and in front of the deep groove, the bones are marked by numerous small pits for the attachment of the muscles of the upper lip.

The portion of the maxilla in which the first premolar is situated, shows that the maxillaries swell considerably on their alveolar border; for the canine then narrow gradually for the first premolar diastema; then swell for its reception, and become very thin in the diastema between the first and second premolars.

The Teeth.—I $\frac{3}{?}$, C. $\frac{1}{?}$, Pm. $\frac{1+}{?}$, M. $\frac{?}{?}$.

The *incisors* have quite deep fangs, are cylindrical in shape, vertical in position, and have no basal ridge.

The crown of the first is worn perfectly smooth, flat across the top, unlike the wearing exhibited in any modern form that has come under our notice. The others have their crowns broken, but indicate a nearly subequal series, with the last slightly the largest.

The canines have long, recurved compressed fangs, with rather short crowns, which are circular in section. They are much larger than in *Procamelus* or *Protolabis* of Pliocene, or the modern camels. (See X., Figs. 3, 5, 6, 7.)

The first premolar is the only one of the molar series preserved ; its crown is chipped considerably, but nevertheless shows that it had one fang and no basal ridge. It resembles the incisors in form, but is slightly narrower and longer.

The genus *Ithygrammodon* has been placed for the present under the *genera incertæ sedis;* but its chief features point to an unmistakable affinity with the ruminants. It is probable that *Ithygrammodon* was the representative and the ancestor, in the eocene age, of that type of ungulates of which the camel and llama are the modern forms.

The line of descent of the camels was first indicated in the Proceedings of the Philadelphia Academy, 1875, p. 262; then in the Bulletin No. 1, 1874, p. 25, of the U.S. Geological Surveys of the Territories, (Government Report of Explorations of 1873, pp. 498-500); and lastly in U. S. Geographical Surveys of Territories, vol. iv., pp. 341-44.

In the above writings, Professor Cope traces the development of the modern camels from the miocene genus *Poëbrotherium*, Leidy, showing the modifications found in *Procamelus*, Leidy, and *Protolabis*, Cope, of the succeeding age.

From the last work (Wheeler's Survey, vol. iv., p. 342), we will quote at length :

"The evolution of the existing types of Camelidæ is a good illustration of the operation of the laws of acceleration and retardation. In evidence of this we may follow the growth of the foot and dentition of the most specialized, and therefore the terminal genus of the series, the American *Auchenia* ——. It is well known since the time of Goodsir, that the embryos of ruminants exhibit a series of superior incisor teeth which disappear early. It is probable, but not certain, that in the miocene genus *Poëbrotherium*, as in various contemporary selenodont artiodactyla, the superior incisors persisted. I have, however, discovered that these teeth persisted in the Loup-Fork genus *Protolabis* during adult life. I have found that one (the second) of these teeth in *Procamelus occidentalis* persisted, without being protruded from the alveolus, until nearly adult age.

"In genera (*e.g.*, the bunodont artiodactyla) where the incisors are normally developed, they appear at about the same time with the other teeth, and continue to develop to functional completeness. This development is retarded in *Protolabis*, since they are not so matured as to remain fixed throughout life in their alveoli.

"In *Procamelus*, the retardation is still greater, since the first incisor reaches very small dimensions, and is, with its alveolus, early removed; while the second incisor only grows large enough, and for a sufficient time, to occupy a shallow alveolus, without extending beyond it. In the first incisor the process of retardation has reached its necessary termination, *i.e.*, atrophy or extinction; while in the existing *Camelidæ* the second incisor also has disappeared the same way. In ruminants other than *Camelidæ*, the third or external incisor has undergone the same process, while in the *Bovidæ* the canines also have been retarded in development, down to atrophy."

Professor Cope continuing the discussion of the teeth of these types, generalizes: "From these and many analogous cases, the general law may be deduced that *identical modifications of structure, constituting evolution of types, have supervened on distinct lines of descent?*"

This summary of Professor Cope's conclusions has been necessary to show clearly in this series the probable place of the eocene genus.

From the shape of the premaxillaries and maxillaries, and the form and position of the teeth the camelline affinities have been indicated. Future research must be relied on to throw light upon the structure of the limbs and the axial skeleton; as it is from these that the conclusive proof must be deduced that *Ithygrammodon* is the ancestor of the camels, and that *Poëbrotherium* is the second link in the chain, instead of the progenitor.

Measurements of *Ithygrammodon cameloides* with *Protolabis*.

	Ithygrammodon c.	Protolabis.
Length of alveolar border of premaxillaries.....	·035	·049
Greatest height of premaxillaries..............	·017	·022
Greatest width of premaxillaries at last incisor...	·013	·0115
Thickness of premaxillary process.............	·010	·017
Length of diastema between first and second incisors.................................	·0035	
Length of diastema between second and third incisors...............................	·003	
Length of diastema between third incisors and canine................................	·004	
Teeth.		
Fore-and-aft diameter of first incisor...........	·007	
Fore-and-aft diameter of second incisor	·0075	
Fore-and-aft diameter of third incisor..........	·008	011
Fore-and-aft diameter of canine...............	·012	012
Fore-and-aft diameter of first premolar.........	·011	·011
Transverse diameter of first incisor............	·007	
Transverse diameter of second incisor..........	·0073	
Transverse diameter of third incisor...........	·0085	
Transverse diameter of canine.................	·0113	·009
Transverse diameter of first premolar..........	·005	.007

AMBLYPODA.

(Cope. Wheeler's Survey. vol. iv., pp. 178 *et ss.*)

"Mammalia, with small cerebral hemispheres which leave the olfactory lobes and cerebellum exposed. The feet short and plantigrade, with numerous (in the known genera, five) digits, terminating in flat, hoof-bearing, ungual phalanges. The seven bones of the carpus distinct, the unciform articulating with both lunar and cuneiform. The astragalus flat, without trochlear surface, and attached to the tibia with very little freedom of movement; its distal extremity divided into two facets, one for the navicular, and the other more or less for the cuboid. Molars inserted with enamel, with wide crowns and transverse crests. A postglenoid process."

This order falls naturally into two sub-orders:

"I. A third trochanter on the femur, and a fossa for the round ligament: no alisphenoid canal; superior incisors present........................*Pantodonta*.
"II. No third trochanter, nor fossa for the round ligament; an alisphenoid canal; no superior incisors.*Dinocerata*."

This sub-order, Dinocerata, includes at present three distinct genera, *Uintatherium*, Leidy, *Dinoceras*, Marsh, and *Loxolophodon*, Cope (also probably *Megacerops*, Leidy). These three genera, in addition to the characters above given, are marked by the possession of two or more osseous projections from the upper surface of the head; and of these the posterior pair, developed from the parietals, are the largest.

Synopsis of genera of

DINOCERATA.

A. Cervical vertebræ long; median horn-like processes anterior to the orbit; nasal tuberosities do not overhang the nasal tips.

> (*a*) Last molar, with or without a tubercle, occupying the entrance of the valley between the lobes, and but one on posterior basal ridge. Temporal fossæ not continued beyond the base of the parietal processes. Occipital condyles projecting...................................*Uintatherium*.
> (*b*) Last molar never has a tubercle at entrance of valley, and has two on posterior basal ridge. Temporal fossæ continued very far back; condyles sessile..............................*Dinoceras*.

B. Cervical vertebræ short; median horn-like processes directly over the orbit; nasal tuberosities overhang the entire tips.... ....*Loxolophodon*.

Synopsis of species of

UINTATHERIUM.

A. With a tubercle occupying the entrance of the valley between the lobes of the last upper molar..*U. robustum*.
B. No tubercle at the entrance of the valley.
> (*a*) Nasals divided by a deep groove; slender zygomatic arch; dorsal vertebræ compressed.
> *U. Leidianum*.
> (*b*) No nasal groove; stout zygomatic arch; dorsal vertebræ subcylindrical............*U. princeps*.

UINTATHERIUM.

Leidy, Proc. Ac. Nat. Sc., 1872, p. 169.—Cont. to Ext. Vert. Faun. of Western Territories, p. 93.—Cope, Hayden's Survey, 1872, p. 580, etc.

Skull broader proportionally than in the other genera of the order, ridged and possessing several concavities on the upper surface; zygomas slender and but little curved; temporal fossæ comparatively short and have well-defined superciliary margins. The cervical vertebræ are rather long; the sacrum has four vertebræ; and the tail is quite long, very flat and broad. The tibia has its proximal face divided by a prominent ridge into two parts. The dental formula for the upper jaw is: I. 0, C. 1, M. 6. The molars are small, and increase from first to sixth. The last is much the largest; in it the anterior lobe considerably exceeds the posterior in size, and there may or may not be a tubercle at the entrance of the valley between the lobes; but there is always one developed from the cingulum at the posterior part of the tooth.

UINTATHERIUM LEIDIANUM, *sp. nov.*[*]

Established upon a head and nearly perfect skeleton of one individual, and parts of two more.

In this species, as in most others of the sub-order, the *nasals* are of immense length and thickness; they overhang the anterior nares, and project considerably beyond the premaxillaries; they form more than half of the entire length of the skull, articulating with the frontals somewhat behind the orbits; the median suture is distinct throughout. Instead of having an expanding shovel-shaped forward projection, as in *Loxolophodon*, they narrow from the median osseous projections (horn cores?) anteriorly. Above the muzzle they are strongly curved from side to side, on their upper surface forming a continuous arch; on the under surface of each bone is a deep concavity, separated from its fellow by the sutural ridge, which disappears forwards, near the extremity. On the upper surface of each nasal, near the forward end, is a large osseous tuberosity which is directed forwards and outwards; these processes are much smaller than the corresponding ones in the *Loxolophodon*, while they are larger, of different shape and direction from those in *Dinoceras*. In this species they are divided throughout by a deep median groove, which anteriorly becomes a fissure, and separates the extremities of the nasals completely. In front of these tuberosities, the nasals taper very rapidly, and

---

[*] This species is respectfully dedicated to Dr. Joseph Leidy, of Philadelphia. The specimens upon which it is established were found on Dry Creek plateau.

end in sharp points which project downwards and forwards; this portion of the bones is shorter, sharper, and projects more decidedly downwards than in *Dinoceras*. Above and slightly behind the sockets of the cranium, the nasals and maxillaries give rise to the median pair of osseous projections. These are very large, subtrihedral at the base, and project upwards, outwards, and strongly forwards. They differ from the corresponding processes in *Dinoceras*, in their forward projection, in being longer and more everted, and in their approach to each other at the base. They taper quite regularly from base to tip, but do not come to a point; a section here would be subcircular. They are not so long and are not knobbed at the ends as in *Loxolophodon*. Their inner curve is convex, the outer concave; converging below at an angle of 90°, they are united by a strong ridge, which is raised decidedly above the surface of the nasals. At their posterior insertion they touch the frontals, lachrymals, and malars. It is probable that the nasals send up processes on the internal and posterior side of these median projections; the sutures between these and the maxillary portions are marked by slight ridges.

Frontals. In our specimen it is very difficult to determine the exact relations of these bones. The nasals narrow at their posterior end, and articulate with the frontals by a V-shaped suture which thus encloses them on each side. It is probable that the frontals run to some point between the posterior or parietal projections. If we have discovered the suture, they overlap the parietals, sending up V-shaped processes, which join the sides of the posterior cranial projections, and form a deep concavity with them. They constitute a large part of the interior wall of the orbit, but have no post-orbital processes; the superciliary ridge is very strong, and gradually rises into a large rounded crest, which joins the posterior projection. The frontal eminences are large, and situated immediately over the lachrymals. The upper surface of the frontals is marked by two high ridges, which probably represent the divided parts of the sagittal crest; they rise from the anterior margin of the bones, and are strongest above the frontal eminences. They converge, and then diverging, together forming an X-like curve (without crossing, however), gradually disappear posteriorly. These various ridges form four deep concavities on the upper surface of the skull: (1) The largest already mentioned, between the parietals and the posterior part of the frontals; (2) that marked by the suture between the nasals and frontal, not so large; (3 and 4) two much smaller lateral ridges between the sagittal and superciliary ridges. This great irregularity of the upper surface of the head is, so far as is yet known, peculiar to *Uintatherium*.

Parietals.—The peculiar feature of these bones is the huge pair of projections (the so-called "posterior horn cores") to which they give rise. The outer margin of these processes is nearly straight; the inner margin for its upper two thirds is also straight, the lower third curving in a high strong ridge to meet its fellow. The anterior face is rounded and produced into the long crest of the frontals, while the posterior face is flattened and produced into a similar but shorter and higher crest which joins the supra-occipital. At the base of these processes their greatest diameter is fore and aft, while at the top it is transverse. They project upwards and outwards, but are not curved as in *Dinoceras*, nor are their upper borders so much arched. Their greatest diameter at top is at right angles to the corresponding measurement in this last-named genus. Behind the ridge which connects these projections, the parietals curve sharply upward to a high occipital crest; between which and the projections they form a deep basin whose floor is raised above that formed by the frontals immediately anterior to it. This arrangement differs from that of both *Loxolophodon* and *Dinoceras*, especially from the latter. The temporal fossæ are of great length, deeply concave, but rather low from above downwards, formed almost entirely by the parietals, and bounded posteriorly by a sharp outward curve of these bones. They are of about the same proportionate length, but higher and very much deeper than in *Loxolophodon;* and are not nearly so long as in *Dinoceras*, in which genus the parietals expand far behind the horn-like processes. In our specimen the posterior part of the fossa is pierced by numerous small venous foramina, and corresponds precisely in every way, save that of size, to the specimen figured by Dr. Leidy. (Cont. to Extinct Vert. Fauna, Plate XXVI., Fig. 1.)

The Squamosals are large and heavy, but encroach little upon the temporal fossæ; they are situated directly below the large parietal projections. The glenoid cavity is transverse and straight in this direction, broad and shallow, with no internal process; and the post-glenoid process is long and massive. The zygomatic process is short, stout, high, and trihedral, with a strongly arched upper margin. The articulation with the malar is by a straight, flat face, and anchylosis of the two never takes place. The anterior termination is pointed, and the outer margin is rounded.

The Malars form none of the face. They are long, slender, curved downwards and backwards, and but little outwards. They are longer, straighter, and less curved outwards than in either *Loxolophodon* or *Dinoceras*; they do not present the sharp angle in the lower margin shown in the latter genus, nor are they so extensively overlapped by the zygomatic processes of the squamosal. Posterior to the molar series, the malars are greatly compressed and very slender, but at the junction with the maxillaries they become much wider and thicker. There is no trace of a postorbital process; and the projections from the under surface of the bone at their junction with the squamosals, so prominent in *Dinoceras*, are here rudimentary or absent. As a whole, the zygomatic arch is very long, slender, simple, curved upwards and very slightly outwards, so slightly that it is completely overhung by the superciliary ridge and frontal crest.

The Lachrymals are unusually large, and form the anterior part of the orbit; they encroach considerably upon the face, and articulate with the superciliary ridges above. The lachrymal foramen is very large.

The Maxillaries are of great length, being nearly as long as the nasals. They extend somewhat beyond the last molar; but the suture between them and the pterygoids is very obscure. There is a very long diastema between the canine and molar series, and the lower margin is arched upwards; above this, between the orbit and the socket of the canine, there is a large, deep fossa, but we can discover no infraorbital foramina. The sockets of the canines are very large and prominent, they curve upwards and backwards to the base of the median cranial projections, but these are not excavated to receive them, as is the case in *Dinoceras*. The most peculiar feature of the maxillary bones is the pair of large horn-like projections, to which, in conjunction with the nasals, they give rise. These have already been described, and it only remains to add that they correspond in position to those of *Dinoceras*, and are therefore much further forward than those of *Loxolophodon*.

The palatine plates of the maxillaries are long and very narrow; they are concave transversely, and are separated from each other by a high median ridge. The posterior palatine foramina are small.

The *Palatines* are very short, narrow, and concave, and are separated by a ridge. They are considerably excavated on the posterior border, in this respect differing from *Loxolophodon*.

The *Pterygoids* and the pterygoid plates of the alisphenoid are compressed; the former join the alveolar borders of the maxillary, which are produced somewhat beyond the last molars. The alisphenoid canal is very large, but rather short.

The *Premaxillaries* are of very peculiar shape, somewhat like a *u*, with one side—the lower—the longer. The upper portion articulates with the nasals, running along the narial opening to about three inches from the angle; the free portion is short, slender, and tapering; it is curved downwards and slightly inwards; the premaxillaries do not meet, leaving the incisive foramen unenclosed, and are edentulous. At the end of the upper portion there are prominent processes for the attachment of the muscles of the proboscis, which probably resembled that of the tapir.

The anterior narial opening is very large, but is considerably concealed by the overhanging of the nasals. There is no osseous septum between the nostrils. The posterior nares are much smaller, being especially contracted in width; it is divided above by the *Vomer*, which is very short, and does not reach the palatines. The nasal cavity thus formed is long, straight, and gradually narrows backwards.

Of the *Mandible* we have but a small portion of the right

ramus, comprising the part opposed to the upper canine, from the dental canal downwards. The jaw at this place has a large downward-projecting process, very similar to that figured by Marsh in his plate of *Dinoceras laticeps*. This process curves slightly outward, and has its external side convex in both directions, and its internal side convex fore and aft, concave from above downwards. The posterior mental foramen, which is the only one preserved in our specimen, corresponds in size and position with *Dinoceras laticeps*. The only difference between the two is a very slight one: in *D.* the curve of the posterior margin of the process is convex; in *Uintatherium* it is concave; the angle which the process makes with the jaw is also greater.

Cranial Measurements.

	M.
Length of bony palate	·227
Length of head along the top	·743
Length of nasals	·398
Breadth at nasal tuberosities	·123
Distance between median projections	·385
Breadth of head posterior to median projections	·189
Breadth of head before posterior projections	·228
Breadth of head behind posterior projections	·312
Distance between extremities of posterior projections	·462
Length of nasals to ridge between median projections	234
Length from ridge between median to ridge between posterior projections	·374
Length of zygomatic arch (straight)	·285
Length from angle of nares to end of nasals	·162
Length from angle of nares to end of premaxillaries	·114
Width between tips of premaxillaries	·053
Circumference of median projection at top	·203
Circumference of posterior projection at top	·329
Height of median projections from ridge between them	·174
Height of posterior projections from ridge between them	·239
Length of nasals anterior to nasal tuberosities	·0415

Teeth.—These are peculiar for their small size compared with the bulk of the animal; for their slender fangs, and for the distance through which these are exposed.

Upper Jaw.—Formula: I. $\frac{0}{0}$, C. $\frac{1}{1}$, Pm. $\frac{3}{3}$, M. $\frac{3}{3}$.

Canines.—The sockets of these teeth have been already described: the fangs are very long, somewhat longer than the crown, and the teeth apparently grew from permanent pulps.

The crown is long, compressed and recurved; the posterior margin is sharp, and exhibits some indications of a slight serration.

Premolars.—The first premolar is not preserved in our specimen. The second and third are subequal, and of about the same conformation. They are implanted by three fangs, one internal and two external. The crown is subcircular at the base. The basal ridge is large, and completely surrounds the crown; above this the crown becomes somewhat trihedral; and is composed of two pyramidal lobes, which meet internally and externally, and are divided by a valley. The summits of the lobes form transverse grinding ridges; the anterior one is the lower, and is crescent-shaped, while the posterior is straight; the anterior ridges throughout the series are much the most worn. The basal ridge is not indented on the outer side at the opening of the valley as in *Dinoceras*.

Molars.—The true molars increase in size regularly backwards. The first molar is larger than the last premolar; it is worn down nearly to the basal ridge, but shows a small internal accessory tubercle on the back part of the posterior lobe. The extremities of the lobes rise into points, and their summits are transversely concave. The number and arrangement of the fangs in the first and second molars is the same as that of the premolars; in the last molar there are but two, which are long and wide, and extend the whole breadth of the crown. This tooth is by far the largest of the molar series; it is ovoidal in shape, with the apex at the outer point of the forward lobe; it is proportionally broader than in *Uintatherium robustum*. The basal ridge is strong, and extends around the entire tooth, with some irregularities of outline at the sides. This tooth, like the others, consists of two pyramidal lobes separated by a valley, which in this case is wide and deep. The summit of the anterior lobe is the longer, and extends obliquely across the crown, while the posterior is nearly straight. The free ends and junction of the lobes are prolonged into points, which give a tripodal character to the crown. The outer point of the anterior lobe is the longest of the three, but the difference is not nearly so marked as in *U. robustum;* the other two are of about the same height;

but the external is spinous in character, while the internal is stout and obtuse. The anterior slope of each lobe is steep, while the posterior is long and gentle. As in *U. robustum*, there is a small rounded tubercle on the inner side of the posterior basal ridge; but there is no tubercle occupying the entrance of the triangular valley between the lobes. The enamel of all the molars is smooth.

Teeth Measurements.

	M.
Length of molar series	·148
Length of premolar series	·062
Length of true molar series	·089
Fore-and-aft diameter of second premolar	·022
Transverse diameter of second premolar	·023
Fore-and-aft diameter of third premolar	·022
Transverse diameter of third premolar	·023
Fore-and-aft diameter of first molar	·023
Transverse diameter of first molar	·026
Fore-and-aft diameter of second molar	·028
Transverse diameter of second molar	·031
Fore-and-aft diameter of third molar	·039
Transverse diameter of third molar	·044
Distance between last molars	·058
Distance between first premolars	·049
Distance between first molars	·074
Fore-and-aft diameter of root of canine	·060
Transverse diameter of root of canine	·039
Length of diastema	·072
Distance between canine alveoli	·080

Lower Jaw.—Dr. Leidy has very kindly sent us some of the lower molars of an undescribed *Uintatherium*, which may belong to this species. They consist of the anterior lobe of the last molar, and second and third premolars entire. They all have nearly the same conformation, consisting of three acute pyramidal lobes, of which the anterior is very much the highest and broadest; the median lobe is partly separated from the anterior by a valley which opens outwards, while the valley between the posterior and median lobes passes completely across the tooth. The highest point of the crown is the inner end of the anterior lobe; but the difference between this and the outer end, very great in the molars, becomes slight in the premolars. The inner end has an accessory tubercle just at the lip of the anterior lobe, and there is another on the anterior basal ridge. This latter feature would seem to distinguish it from *U. robustum.*

The basal ridge is distinct in front and behind, indistinct on the external side, and entirely absent from the internal.

Measurements.

(TEETH, LOWER JAW.)

	M.
Transverse diameter of last molar	·026
Height of highest point above cingulum of last molar	·023
Fore-and-aft diameter of second molar	·0285
Transverse diameter of second molar	·023
Height of second molar	·0215
Length of fourth premolar	·021
Width of fourth premolar	·015
Height of fourth premolar	·020
Length of third premolar	·020
Width of third premolar	·015
Height of third premolar	·014

Vertebræ.

Cervical region (Plate VI., Fig. 1).—(Only one preserved, probably fifth or sixth.)

The centrum is short, compared with the dorsals, but is much longer than the cervical centra of the *Proboscidea;* it is broad and depressed, oval in form, and slightly opisthocœlous. The zygapophyses are developed upon tuberous projections of the pedicles; they are large, flat, and in the same plane with each other. The diapophyses are very slender and short, and but slightly heavier than the parapophyses, with which they unite, enclosing a large vertebraterial canal. At the anterior margin of the parapophysis, a small pointed process projects downward.

The pedicles are low and very heavy, bounding a narrow neural canal.

The epiphyses are not so completely ossified as they are in the dorsal region.

Dorso-lumbar region (Plate VI., Figs. 2, 3, 4, 5).—(Description based upon nine dorsals and two lumbar vertebræ.)

The centra are large, subtriangular, and slightly compressed; they are opisthocœlous, but less so than in the *Proboscidea*. They increase in size slowly but regularly from before backwards. In the middle dorsal region they are excessively expanded laterally for the posterior-costal attachments; but become less broad and higher as they recede in the series. In the middle of the series the centra are marked by a prominent hypophysial keel.

In the anterior dorsals the costal surfaces are developed almost entirely upon the pedicles, and are two in number. These are very large in about the sixth and seventh, and meet. They decrease in size as they recede. In the middle region the posterior costal surfaces are small, lozenge-shaped facets, and are developed upon thin lateral projections of the centra.

The centrum of the last dorsal is heavy, resembling the centra of the lumbar, from which it is distinguished by a single pair of small costal surfaces, developed, half on the pedicles and half on the centrum.

The neural spines are markedly smaller than those of the *Proboscidea* and *Rhinoceros*; in the anterior part of the dorsal region the spines have the same angle as the corresponding ones in *Mastodon*. The spines of the twelfth and thirteenth (approximately) are much expanded at the ends and bifid; in the last dorsals they are wide, straight, and very thin; in the lumbar region they are short, stout, tuberous, and stand almost at right angles to the axis of the vertebræ. The laminæ in the anterior part of the series are long and thin, decreasing in length but increasing in thickness from before backwards. The zygapophyses in the fore part of the dorsals are mere flat facets on the laminæ; they increase in size and become characteristic in the posterior part of the dorsal region. The pre-zygapophyses of the last lumbar vertebra are very large, deeply concave, and parallel with the axis of the column. The metapophyses appear in the middle of the series, and regularly increase in size to the last lumbar. From their appearance the pre-zygapophyses are developed upon them.

The transverse processes present the most unique feature of the vertebral column. In the anterior region they are long, wide, and rugose, and in the same plane with the laminæ; they send directly out wide, downward-curving projections.

In the middle dorsal region the transverse processes lose these thin projections, and become heavy and subtrihedral, with a smooth facet on their lower face for articulation with the tubercle of the rib; their upper face is here a little twisted from the plane of the laminæ. In the posterior dorsal region the transverse processes lose the facet for the articulation with the tubercle of the rib, and become short and very thin, pointing slightly backwards.

The lumbar transverse differ from those of the last dorsal in having a median transverse ridge on their posterior face, and are directed more vertically out.

The pedicles throughout most of the dorsal series are short, heavy, and deeply notched behind. In the posterior dorsal region they become longer and more slender, resembling almost exactly the pedicles of the lumbar region.

Sacral region.—The sacrum is composed of four vertebræ, three true and one pseudo-sacral. The centra are extremely depressed, and rapidly decrease in width and thickness from before backwards. The first is shorter than the last lumbar, but much longer than the other sacrals, which are subequal.

The face of the first is elliptical, and is nearly three times the diameter of the fourth, longitudinally. The metapophyses are exceedingly large and tuberous on the first; with wide, deeply concave pre-zygapophyses developed upon them; in the other three the metapophyses are rudimentary.

The transverse processes are long and wide in the first three; widest in the first but thickest in the second; long and thin in the fourth. The pleuropophysial segments of the true sacrals are very heavy.

The foramina enclosed by the transverse processes are large; and have, on the internal side, their long diameter obliquely transverse to the axis of the sacrum.

The inferior faces of the centra are slightly concave in the first three. The first and fourth have slight hypophysial keels.

The neural canal is very wide and depressed in the first, but decreases rapidly backwards, becoming extremely small in the last.

Caudal region (Plate VI., Fig. 6).—(Description based upon the first four.)

The centra of the caudal vertebræ are rather long, narrow, and greatly depressed in the middle; they decrease in size gradually backward. The pedicles and laminæ are short and thin, enclosing a small neural canal. The neural spines point directly back, being almost parallel with the axis of the centrum. They are slender and tuberous at the extremity. The transverse processes are very long, wide, and thin, thickened somewhat at the ends, and project directly out; they decrease in size backwards. From the persistence of the neural canal, and from its comparative size, it would seem as if the tail was considerably larger than that of the elephant.

Measurements of Vertebræ.

Cervical region (Plate VI., Fig. 11).

	M.
Diameter of cervical (vertical)	·079
Diameter of cervical (transverse)	·111
Diameter of cervical (fore and aft)	·051
Extreme length of prolongation of pedicles for zygapophyses	·089
Long diameter of vertebraterial canal	·040
Width of pedicles	·035

Dorsal region (Plate VI., Figs. 2 and 3).

	M.
In anterior region: Diameter of centrum (fore and aft)	·075
Diameter of centrum (vertical)	·062
Diameter of centrum (transverse)	·142
In posterior region (Plate VI., Fig. 3).: Diameter of last dorsal (vertical)	·087
Diameter of last dorsal (transverse)	·110
Average width throughout the series of neural canal	·097
Width of transverse process (anterior dorsal) (Fig. 2, Plate VI.)	·075
Length of lamina from prezygs. to post-prezygs. (Fig. 2, Plate VI.)	·12
Length of transverse process (middle dorsal), (Fig. 3, Plate VI.)	·055
Vertical length of anterior dorsal from tip of spine to hypapophysial keel	·183
Width between inner margins of metapophyses (last dorsal)	·172

Lumbar region—Last lumbar (Plate VI., Figs. 4 and 5).

	M.
Diameter of centrum (fore and aft)	·095
Diameter of posterior face (transverse)	·111
Diameter of posterior face (vertical)	·078
Extreme width between transverse processes	·214
Width between pre-zygapophyses	·110
Length of neural spine from lamina	·060
Vertical length from tip of spine to lower side of centrum	·194

Sacral region.

	M.
Length of sacral series	·264
Transverse extent of sacral series	·298
Diameter of first vertebra at free end (transverse)	·109
Diameter of first vertebra at free end (vertical)	·072
Diameter of first vertebra at free end (fore and aft)	·070
Diameter of last vertebra at free end (transverse)	·053
Diameter of last vertebra at free end (vertical)	·031
Diameter of neural canal at first vertebra (transverse)	·100
Diameter of neural canal at last vertebra (vertical)	·018
Long diameter of first transverse foramen (internal side)	·076

Caudal region (Fig. 6, Plate VI.).

	M.
Length of first four	·252
Diameter of centrum, first (fore and aft)	·061
Diameter of neural canal of first (transverse)	·035
Extent of transverse process in first	·085
Extent of transverse process in fourth	·046
Fore-and-aft width of transverse process of first in the middle	·048

Ribs.—Description based upon two perfect ones, and parts of four more.

The ribs, as in *Dinoceras*, "resemble very much those of the *Mastodon*." The capitulum has two convex facets, separated by a narrow groove. The sternal end in one is very much expanded, but less so in the other. The tubercle is small, and is situated upon the prolongation of the lamelliform process for the intercostal muscles. The angle is much sharper than in the corresponding ribs of *Mastodon*.

Measurements of Ribs.

	M.
Length of rib without curvature	·571
Length of rib with curvature	·739
Width of rib just below head	·049
Width at muscular attachment	·063
Vertical diameter of larger facet of capitulum	·039

Scapula (Plate VIII., Fig. 1).—The scapula is sub-triangular, with the pre-scapular border as base, and the apex half way up the post-scapular.

On the external side the pre-scapular fossa is concave antero-posteriorly, and plane in the direction of its length; it is very thin in the middle, but becomes heavier at the lateral border. The postscapular fossa is larger and less concave; both fossæ, as they approach the upper border, gain greatly in thickness.

The spine rises from the supra-scapular border, and extends to within an inch from the glenoid cavity. It is decidedly antroverted; thickest and highest near glenoid cavity; thinnest and lowest in the middle, expanding again at its upper extremity; its acromio-scapular notch is long and shallow. The acromion is rudimentary.

The coracoid process is a low, rugose tuberosity; the coraco-scapular notch is short and low.

The internal surface of the scapula has a large, smooth median ridge, extending the whole length of the bone, separating it into slightly concave fossæ. About two thirds up, the median ridge sends obliquely up two branches, forming a V, superimposed upon the main ridge; but these disappear before they reach the upper extremity. The whole inner surface is curved outward, presenting a concave appearance throughout.

The glenoid cavity is deep, ovoid in form, with its greatest diameter fore and aft, and its smaller end behind.

The resemblance between the scapula of *Uintatherium* and that of the *Proboscidea* is more closely marked than in any other corresponding bones. The chief points of similarity are: first, the subtriangular shape; second, the same relative proportions between the fossæ; third, the antroversion of the spine; fourth, the glenoid cavity looking directly down. The marked points of difference may be summed up as, first, in the dissimilar proportion of the glenoid cavity; second, the great thickening of the spine at its upper and lower extremities; third, in the high, shallow acromio-scapular notch; fourth, in the longitudinal concavity of the internal surface.

Measurements.

	M.
Extreme length of scapula	·42
Extreme length of spine	·31
Extreme thickness of spine at upper margin	·098
Extreme height of spine	·109
Basal width of proximal end of spine	·032

Basal width of distal end of spine............................ ·030
Basal width of middle part of spine........................... ·011
Length of glenoid cavity..................................... ·149
Width of glenoid cavity...................................... ·096

The Humerus (Plate VII., Fig. 1).—The humerus is short, but excessively stout, twisted slightly on its axis; it decreases gradually in size downward, with the smallest diameter about two inches below the end of the deltoid ridge. The head is large, hemispherical, and sessile, projecting very slightly out of the axis of its shaft. The great tuberosity is heavy, but not high, and is separated from the low lesser tuberosity by a shallow bicipital groove.

The trochleæ are very nearly equal in size, directed obliquely to the axis of the shaft, and are separated by a narrow groove, which runs from the supra-trochlear fossa down and in, then back and up to the anconeal fossa. The condylar tuberosities are large and rugose; the external is the greater, and is directed antero-posteriorly. The deltoid ridge is long and heavy, and extends nearly two thirds down the shaft, branching out into two forks near its end. The supinator ridge is short and rudimentary, differing in this respect entirely from the great development found on the humerus of the *Proboscidea*.

The supra-condylar fossa is small, sub-circular in form, and very deep, it has the peculiarity of being placed above the external condyle alone. The anconeal fossa is median in position and quite deep.

Measurements of Humerus.

	M.
Length...	·63
Smallest circumference of shaft just below deltoid ridge...............	·258
Greatest proximal circumference below greater tuberosities............	·45
Width of trochlea on anterior side........................	·154
Length of anconeal fossa.................................	·071
Width of anconeal fossa..................................	·062
Width of distal end at condylar tuberosities....................	·23
Length of deltoid ridge...................................	·22
Length of groove running from supra-trochlear to anconeal fossa........	·272

The Ulna (Plate VII., Fig. 2).—The ulna is long, heavy at both ends, with a slender trihedral shaft that curves forward, and decreases in size as it approaches the distal end.

The proximal end has its articular face for the humerus divided into three facets, the vertical being long and extremely convex; the horizontal has the pre-axial face the longer. The olecranon is very massive and rugose, sending up on its external side a high, pointed projection; while on the internal side, the olecranon becomes compressed and projects inward; these tuberosities are separated by a wide, shallow groove for the tendon. The distal articular face is large, single, and is concave antero-posteriorly, and convex laterally. The styloid process is very heavy.

Measurements of Ulna.

	M.
Length of sigmoid notch	.089
Width of sigmoid notch (horizontal face)	.049
Diameter of proximal end below articular face (fore and aft)	.076
Diameter of proximal end below articular face (transverse)	.059
Diameter of distal articular face (fore and aft)	.061
Diameter of distal articular face (transverse)	.051
Length of olecranon	.112

Pelvis.—The *ilia* are greatly expanded laterally, with the iliac surface concave and the gluteal surface nearly flat; thin in the middle, they increase in thickness near the borders. The crests curve regularly, and project above and beyond the sacrum, but do not bend over the acetabula. The acetabular borders are only slightly concave; the prominence for the attachment of the rectus muscle (extensor) is low and V-shaped. The direct internal surfaces of the ilia, comprised between the pubic and ischiatic borders, become deeply concave below the ischiatic portion of the acetabulum. The sacral surfaces are wide and triangular in shape; above they project beyond the sacral spines.

The ilio-lumbar angle is about 110°. The ischium is short; a section of it, as it leaves the acetabulum, is subtriangular; but immediately it becomes flattened fore and aft throughout its plane of 90° to the axis of ilium; then verging toward the pubic symphysis it becomes small. The tuberosity of the ischium is small and directed up.

The pubis, as it leaves the acetabulum, is sub-cylindrical, after this it is flattened in the same plane with the ischium

The bone as a whole is short, thin, and slightly curved on its own axis; its smallest part makes up the pubic symphysis, which is short. The thyroid foramen is a large oval, with its long diameter parallel to the axis of the ischium.

The acetabulum is large, sub-circular, and deep, with prominent borders; especially the iliac, which is produced on its external extremity into a point; the ischiatic is deeply notched. From the wide ligamentous pit in the centre there runs a deep groove part way down the antero-external side of the ischium.

The anterior opening of the pelvis is a wide oval, with its longest diameter transverse.

Measurements of Pelvis.

	M.
Transverse diameter of pelvis, including sacrum	1·171
Long diameter of ilium (from lower margin of the crest to sacral surface)	·440
Short diameter of ilium (from acetabulum to upper margin of crest)	·393
Length of acetabular border	·125
Length of ischium	·220
Width of ischium at tuberosity	·124
Length of pubis	·196
Greatest width of pubis	·061
Smallest width of pubis	·027
Long diameter of acetabulum	·139
Short diameter of acetabulum	·118
Long diameter of thyroid foramen	·094

The Femur (Plate VIII., Fig. 4).—The femur is short, with a small oval head, strongly compressed fore and aft. It is less out of the axis of shaft than in *Proboscidea*, and has no pit for the ligamentum teres. The shaft is straight and simple, much compressed transversely at the proximal extremity, becoming sub-cylindrical below.

The great trochanter is heavy, rugose, and strongly recurved; it is separated from the shaft by a wide and deep digital fossa. The second trochanter is a mere rudimental tuberosity. The condyles are nearly of an equal size, very convex, and are divided by a deep popliteal groove.

The condylar tuberosities are low, the internal sends obliquely a ridge three inches long, up and across the axis of the shaft at an angle of 45°, that forms the upper boundary of the popliteal space.

The front part of the trochlear faces for the patella are gone in our specimen.

Measurements of Femur.

	M.
Greatest distal diameter across condyles	·178
Length of condyles	·079
Transverse diameter of shaft six inches above distal end	·121

The Tibia (Plate VIII., Fig. 2).—The tibia is short, straight, and simple, slender in the middle, but much expanded at the extremities. The proximal end is especially massive, with deeply concave articular faces; the internal cotylus is the larger, and has its greatest diameter fore and aft, directly at right angles to the greatest diameter of the external.

The cotyli are separated by a smooth ridge, that is highest at its posterior termination. The tuberosity is high and massive, with a wide depression on its top for the reception of the ligament of the patella; the sides are deeply pitted by venous foramina. A section of the shaft at the lower part of the tuberosity would be subtriangular; while below it becomes more cylindrical, as it becomes smaller, reaching its shortest diameter about three fifths down the shaft. Below this it expands and forms the large subcircular distal end. The articular face is concave, with a slight, smooth ridge runing fore and aft on its internal side. The malleolus is broad and low.

The *fibula* is distinct, but very slender; its proximal end has a small circular face, which articulates with the tibia on a facet developed on the bottom of a projection of the tuberosity; the distal end is much flattened, but enters into the ankle-joint.

Measurements of the Tibia.

	M.
Transverse diameter of proximal articulation	·158
Longitudinal diameter of proximal articulation	·123
Smallest circumference of shaft	·16
Greatest distal circumference above the articular face	·308
Diameter of distal articulation longitudinally	·082
Diameter of distal articulation transversely	·130
Width of tuberosity below pit for ligament of the patella	·075

UINTATHERIUM PRINCEPS, *sp. nov.*

Established on several portions of the head, vertebral column, and the limbs.

This species may readily be distinguished from the last by its larger size, the broad nasals with small tuberosities, the stronger zygomatic arches, and the sub-cylindrical centra of the dorsal vertebræ.

The *nasals* are broad and flat for some distance behind the tuberosities. Although the animal was not adult, the nasal suture is almost obliterated, and is visible only on the inferior surface; in this respect it is very different from the *U. leidianum*, in which the suture persists throughout life; as is shown by the type specimen, which was past maturity. The tuberosities of *U. princeps* are lower, broader, and more everted than in the preceding species; are not divided by a groove, but united together by a low rounded ridge, to which the surface of the nasals gradually rises from behind. The portion of the bones anterior is short, stout, and projects horizontally.

The *median*, or *maxillary*, *projections* are apparently short and everted; they are somewhat compressed antero-posteriorly, and enlarge rapidly downwards. They are transversely oval in section.

The *posterior*, or *parietal*, *projections* are different from any that we have yet seen. They are sub-trihedral at base and flattened antero-posteriorly above. The upper margin is regularly arched, and is the thinnest portion of the process. The posterior face is perfectly flat, and the parietal crest makes a sharp angle with it, rising lower down than in *U. leidianum;* the frontal crest is also lower down than in that species, and the anterior face is strongly convex. The internal margin is rounded and straight, and does not show the longitudinal groove marked in the last species.

The *frontal* has the eminences and depressions common to the members of the genus; but the superciliary ridge is unusually sharp and sinuous in outline.

The *squamosal* is short, stout, and high; is but slightly curved outwards, but apparently projects somewhat downwards. The malar articulation is broad and flat, indicating the heaviness of that bone. As a whole the zygomatic arch is stronger, and probably shorter, than in *U. leidianum*.

The occipital condyles are proportionately rather small; they are placed on a long neck and project downwards. They are strongly convex from above downwards, but scarcely at all so from side to side. The internal border is slightly emarginate.

The *dorsal vertebræ* are of about the same proportionate length as in *U. leidianum*, but are higher and wider; and the centra are sub-circular in section, slightly contracted in the middle. The costal surfaces are wide and deep, and vertically oval in shape; they are placed partly on the centra and partly on the neurapophyses. The transverse processes are short, stout, and tuberous, and raised high above the centrum. There is a deep notch at the posterior edge of this process, at its junction with the neurapophysis. The neurapophyses are trihedral, somewhat low, and very stout, forming a wide neural canal.

The *ulna* (Plate VII. Fig. 2) is thick, with a long and rugose olecranon. The shaft is long and stout; it shows a distinct medullary cavity. The distal end is small, and shows a low, heavy, styloid process.

A *metacarpal* resembles the corresponding bone in *Dinoceras*, but is rounder and less rugose. There are two faces for carpal articulations, which meet at an open angle.

Measurements.

	M.
Breadth over nasal tuberosities	·132
Length of nasal tip (anterior to tuberosities)	·033
Circumference of maxillary projection near the tip	·208
Vertical diameter of squamosal	·058
Transverse diameter of ditto at malar articulation	·045
Length of centrum of dorsal vertebræ	·070
Height of ditto	·083
Breadth of ditto below costal surfaces	·097
Breadth of pedicle	·058
Length of transverse process	·056
Extreme breadth of transverse process	·061

Found at Henry's Fork.

UINTATHERIUM ROBUSTUM. Leidy.
Cont. to Ext. Vert. Faun., p. 96.

A fragmentary skeleton from Henry's Fork.

RODENTIA.

PARAMYS, Leidy.

"Extinct Vertebrate Fauna of the Western Territories," vol. i. Hayden's Surveys, p. 109 *et seq.*

This genus of the gnawers is very closely allied to the squirrels and marmots. Dr. Leidy sums up the generic characteristics of the teeth thus:

"The four lower molars are proportionately narrower than in squirrels and marmots, the fore and aft exceeding the transverse. The crowns are short, square, tuberculate, and enamelled.

"The lower jaw is proportionately shorter and deeper than in most known rodents; the reduction in length being mainly due to a less development of that part of the bone in advance of the molars. To compensate for the difference in length, and to make room to accommodate the incisors, these teeth reach further back than usual.

"The acute edge of the hiatus between the molars and incisors is almost on a level with the alveoli, of the teeth, instead of forming a deep concave notch, so conspicuous a feature in the jaws of the gnawers generally."

Species known.—

Loc. cit. { PARAMYS DELICATUS, Leidy.
PARAMYS DELICATIOR, Leidy.
PARAMYS DELICATISSIMUS, Leidy.

PARAMYS ROBUSTUS, Marsh.

Am. Journ. Sc. v. iv., p. 218

PARAMYS SUPERBUS, *sp. nov.*

Established on a single lower incisor.

The species indicated by this specimen is the largest of the genus as yet known. The tooth is subtrihedral in section, is stout, and shows but little curvature. The anterior and lateral faces are broad, and the under margin to which the sides converge, is narrow. The enamel is thick and is inflected so as to cover a small portion of the sides as well as the front.

Measurements.

Transverse diameter of crown	·0062
Fore and aft diameter of crown	·007

Found at Cottonwood Creek.

AVES.

THE expedition collected parts of four species of birds; a feather from Florissant Col. the distal ends of a humerus, and a femur and a portion of the shaft of an ulna (?) from the Bridger beds, but they are all too uncharacteristic for classification or description.

REPTILIA.

CROCODILIA.

CROCODILUS.

CROCODILUS APTUS, Leidy.
Contributions to Extinct vertebrate fauna of the Western Territories, p. 126.

Henry's Fork.

CROCODILUS GRINELLI, Marsh.
American Journal of Science and Arts, vol. i., p. 465.

From Cottonwood Creek.

CROCODILUS ELLIOTII, Leidy.
Cont. to Ex. Vert. Fauna, p. 126.

Represented by a perfect skull, and several vertebræ.

This fossil exhibits a form of skull which shows characters of both crocodile and alligator; the latter to a less marked degree. That it properly belongs to the former genus is shown by the notch in the upper jaw which receives the canine of the lower.

The entire skull is remarkably flat on its upper surface, the face and cranium being nearly in the same plane without the descent at the frontals usual in these reptiles. The jaw is deeply notched at the sutures between the maxillaries and pre-maxillaries, and the second maxillary notch is well marked. All the bones of the upper surface of the head are deeply pitted.

The borders of the cranium are rounded as they approach the orbits; the superior temporal orifices are almost perfectly circular, the fore-and-aft diameter exceeding the transverse by only one millimetre. This effect may, in some degree, be due to distortion.

To give a more detailed account of the several elements of the skull, we take up first the *basioccipital*. This bone is remarkably long and straight, tapers gradually downwards, and becomes quite narrow at the distal end. It is smooth throughout, and exhibits no rugose muscular attachments, such as are sometimes seen in other members of the order. The condyle is large and nearly spherical, but with median groove distinctly marked. In size and shape it is more like that of the alligator than of the ordinary crocodile, but it is somewhat different from either. It differs from the former, in not having so long a neck distinctly marked by a constriction; and from the latter, in not having additional articular faces on each side of the condyle proper. As far as can be judged, no portion of it is formed by the exoccipitals. Below the condyle, the basioccipital is perforated by two small vascular foramina; the spheno occipital canal occupies the usual place, and is very large.

The *exoccipitals* are large, of very great width, but rather low from above downwards. The position of the foramina which perforate these bones is peculiar; it resembles more the arrangement seen in the skull of the alligator than in that of the crocodile, but it has an additional foramen. There are, then, two small venous foramina near the condyle; while along the lateral margin of the occiput, are placed in a vertical line the foramina for the hypoglossal and pneumogastric nerves, and the internal carotid artery. The foramen for the facial nerve, etc., is situated in the usual place, and is of the usual size. The foramen magnum is heart-shaped, low, wide above, contracting below. The paroccipital processes are long and slender, and project strongly backwards.

The *supraoccipital* is very small. It shows to some extent on the upper surface of the skull, wedged in between the parietals. As a whole, the occiput is of remarkable shape: it is perfectly vertical, as in all other crocodilians; is remarkably high from above downwards, and is very broad at top, becoming extremely narrow distally. This latter feature is owing to the peculiar shape of the pterygoids; which, when viewed from behind, do not appear to reach the basioccipital.

The *basisphenoid* is so destroyed by crushing, that nothing can be said of its shape or of its foramina. It was, however, evidently very stout. The alisphenoid is large and smooth and is not ridged; the *foramen ovale* is small, and is not produced into an anterior notch. The suture between the alisphenoid and the pro-otic are quite distinct, but the latter is so firmly anchylosed to the quadrate that its limits are indeterminable. None of the other periotic bones are visible.

The *quadrate* is very long and broad. In shape it resembles the corresponding bone of the alligator, but is somewhat broader. The lower surface is divided into two unequal parts by a conspicuous ridge, which runs to within an inch of the articular surface. This surface has a more decidedly grooved or trochlear appearance than in the common crocodile; but not to the same degree as in the specimen described by Dr. Leidy. It is more like the Mississippi alligator in this respect than any living species of which we have specimens.

The *pterygoids* are of peculiar shape; they are long, slender, and pointed, and meeting the basioccipital on the median line in front, they project downwards and backwards. Their posterior border is very deeply emarginate, so that they seem to have no connection with each other, or with the basioccipital; instead of having the broad, plate-like appearance of these bones in recent species. The suture between these on the palatal surface is long; and the processes of the pterygoids, which bound the posterior nares, are long and stout. The posterior nares have the position which they take in the recent species; they are visible in the occipital surface, and are directed backwards as well as downwards. They are rather small, and appear to have no septum between them, but this cannot be said with any certainty.

The *transpalatine* is also somewhat peculiar in shape; the process which joins the pterygoid is of great length, being nearly as long as that bone. The other limbs are more normal in length. The three processes are connected at the usual angle.

The *palatals* are long and narrow, becoming wider anterior to the foramina. The suture with the maxillaries is rounded, and there are no forward processes as in the true crocodiles; but at the same time, these bones are not of the shape exhibited in the alligators. The palatal foramina are of immense length; they are more than one third as long as the entire bony palate. Their width is also considerable.

The *maxillaries* are long and very broad; the alveolar border is of about the same shape as in the crocodile, but less decidedly sinuous; and the posterior part passes in below the alveolus of the lower jaw. The convexity of the upper surface of the maxillaries is not so well marked as in the true crocodiles. It is nearly as flat as in the alligator. The palatine plates of the maxillaries are short, broad, and nearly flat, arching slightly to form the alveolus, and perforated along this border by rows of foramina.

The *premaxillaries* are very short; they curve strongly outwards from the notch and enclose the large anterior narial opening, which is distinctively *crocodilian* (as distinguished from other genera) in shape. The muzzle ends quite sharply. The palatine processes are short and *convex* in both directions, and the incisive foramen is heart-shaped. The alveolus is quite regular in outline; it is pitted in front for the first mandibular teeth, but there is no perforation for them.

The teeth are short, stout, compressed so as to form cutting-edges, and are somewhat obtuse. They are finely striated from base to tip. The premaxillary held four teeth; these are all broken off, but from their fangs it appears that they formed an uninterrupted row, and were subequal in size. The maxillary accommodated fifteen teeth, of which the fifth is the largest, and forms a very prominent canine. The posterior maxillary teeth are proportionately larger, and more equal than in either crocodile or alligator.

The *nasals* are rather broad; they send processes into the anterior narial opening, which tend to divide it, but this division was probably not complete. The limits of the *prefrontals* and *lachrymals* are so obscure as to preclude description.

The *frontals* are long and very narrow, though wider than in the alligator. They are smooth and flat, exhibiting no concavity on top. They expand considerably at the posterior part of the orbit, at the sutures with the post-frontals. These bones are long, stout, and curved very strongly outwards.

The *mastoids* are of remarkable size; they project far backward along the tympanic, and encroach largely upon the occipital region. It is partly owing to this that the occiput is so high.

The *parietals* are short, and very narrow between the temporal orifices, behind these they expand considerably. The orbits are large and of irregular shape.

The *malar* is long and rather slender.

Measurements.

	M.
Length from occipital border to end of muzzle	·455
Breadth of cranium at postorbital angles	·0935
Breadth of cranium between temporal orifices	·019
Breadth of forehead between orbits	·036
Breadth of temporal orifices	·037
Fore-and-aft diameter of the same	·038
Fore-and-aft diameter of the orbits	·073
Length of face in advance of the orbits	·3055
Breadth of face outside of the fifth maxillary tooth	·176
Breadth of muzzle as formed by premaxillaries (about)	·114
Breadth of muzzle at notch for canine	·093
Length of premaxillaries to notch	·080
Estimated length of entire alveolar border	·283
Breadth of articular surface of quadrate	·059
Vertical height of occiput	·124
Vertical diameter of foramen magnum	·016
Transverse diameter of foramen magnum	·025
Vertical diameter of condyle	·023
Transverse diameter of condyle	·028
Length from palatine foramen to end of pterygoid	·120
Length of palatals	·1295
Length of palatine foramen	·142
Greatest breadth of palatine foramen	·051
Length of bony palate from incisive foramen to posterior nares	·350

The *mandible* is long, rather shallow, but very thick. The symphysis is very long, extending as far as the seventh tooth. The chin is quite sharp, but becomes broad, as the rami diverge quite rapidly. The alveolar border is rounded and comparatively straight, the median enlargement of the dentary is in thickness rather than in height. The two rami diverge at an angle which is more open than in the true crocodiles, and less so than in the alligator. The mandibular foramen is smaller than in the latter genus, but corresponds with it in position; it has its long diameter parallel with, and not oblique to, the alveolus. The mandibular fossa is extremely large and deep. The splenial is long and stout, ends obtusely, and does not reach the symphysis. The articular cavity is broad from side to side, but rather shallow; it is not divided into two distinct facets as in the alligator. The post-glenoid process is very stout; it projects but slightly upwards, less so than in either crocodile or alligator. It has no median ridge as in the former genus, and is tuberous at the end.

The teeth are much like those of the upper jaw, but are somewhat sharper and more conical; they are not recurved, and have no distinct constriction, as is found in most of the recent species. There are about eighteen teeth to each ramus; the first is large and sharp, and is followed by two small ones and then by the canine. The remaining teeth do not exhibit any great differences of size.

Measurements.

	M.
Length of rami (straight)	·618
Width of lower jaw outside of glenoid cavities	·316
Length of symphysis	·110
Width of jaw at second enlargement	·057
Depth at oval foramen	·073
Greatest width at symphysis	·110
Space occupied by teeth	·340
Breadth of glenoid cavity	·072
Length of post-glenoid	·065

Vertebræ.

Cervicals.—The centra are long and nearly cylindrical, expanding slightly near the anterior face. The hypapophyses are short and broad; they project forwards, and are somewhat compressed. The vascular foramina correspond in size and position to those of the alligator. The neural canal is small and subcircular, having comparatively short but very stout neurapophyses, which are perforated behind the diapophyses. These are short and stout, and are developed from the neurapophyses alone. The zygapophyses are long and flat; the anterior ones project almost vertically. The neural spines are stout, and of greater antero-posterior extent than in the alligator.

Dorsals.—A few of the anterior dorsals have strong hypapophyses. The centra of all are long and stout. The neural canal is smaller than in the cervical region; it has strong neurapophyses, which develop long and depressed diapophyses. The neural spines are low and broad. The zygapophyses are developed on the laminæ, and do not project upwards as in the cervical region.

The *lumbars* are much like the dorsals, except that they are longer, and have very large and depressed diapophyses, which project somewhat upwards.

Measurements.

	M.
Length of centrum of a posterior cervical	·049
Height of neural canal	·015
Fore-and-aft diameter of neural spine	·023
Length of diapophysis	·023

Dorsal.

	M.
Length of centrum	·0555
Length of hypapophysis	·018
Fore-and-aft diameter of neural spine	·030

Lumbar.

	M.
Length of centrum	·057
Fore-and-aft diameter of neural spine	·0285
Diameter of diapophysis	·029

The dermal scutes are long ellipses, deeply pitted on one side, but without a trace of a keel. Their edges show no signs of sutural union.

The bones described indicate a reptile about fifteen feet long. They were found near Smith's Fork, Wyoming.

CROCODILUS PARVUS, *sp. nov*.

A small reptile represented by sixteen vertebræ and a portion of the pelvis.

The *cervicals* have short centra, with very deep articular cups and hemispherical heads: the latter have a prominent rim around the base. The hypapophyses are short, stout, and very nearly vertical in direction; the sides of the centra are channelled by a deep vertebraterial canal; the parapophyses are developed very low down, and are very prominent. The diapophyses are developed partly from the centra and partly from the neurapophyses; they are stout and very short. The facets for the ribs are developed in the usual place. The neural arch is rather high, and forms a small, narrow canal; the proportions of the neurapophyses and neural spines are about as in *Alligator mississippiensis*, but the zygapophyses are not so prominent.

The *dorsals* show a considerable increase in size over the cervicals; the centra become elongate, and the articular cups shallower and transversely oval. Several of the anterior vertebræ retain large hypapophyses. In the dorsal region the diapophyses are developed from the neurapophyses alone, and at a considerable height above the neuro-central suture; they are long and depressed. The zygapophyses and neural spines present no peculiarities of structure.

In the *lumbar* region the vertebræ regain their cylindrical form, and become still more elongate; but the cups are comparatively shallow and the heads low; there is no distinct shoulder. The neural canal is small, with low, broad neurapophyses, from which are developed very broad and depressed diapophyses, which project outwards in a horizontal plane. The neural spines are thick, and broad antero-posteriorly, but are so broken that their height cannot be determined.

In both dorsal and lumbar region the neurapophyses are deeply notched on their posterior edges, and close to the neuro-central sutures, for the passsage of the spinal nerves.

The *ilium* is very high compared with most crocodiles; its vertical diameter is considerably more than half of the antero-posterior diameter. The construction of the bone is very much like that in the modern *Crocodilia*; but the suprailiac border is more regular, and the anterior tuberosity is not so thick or so much everted as in these forms. The acetabulum is small, shallow, directed downwards and outwards, and situated considerably forward of the median line. The iliac surface is smooth, but very irregular, being very deeply concave above the acetabulum, and convex behind it. The sacral surface is rugose, rises above the sacrum, and shows attachments for two sacral vertebræ. Thus the construction of the entire pelvis shows but very little variation from the modern type.

Measurements.

	M.
Antero-posterior diameter of ilium	·095
Vertical diameter of ilium	·061
Antero-posterior diameter of acetabulum	·029
Length of centrum of third (?) cervical vertebra (from edge of cup to tip of ball)	·032
Height of neural canal, third (?) cervical	·0095
Length of hypapophysis of third cervical	·0085
Length of centrum of a posterior dorsal	·040
Length of diapophysis (about)	·032
Length of centrum of a lumbar	·046
Vertical diameter of a lumbar	·029
Breadth of neurapophysis of a lumbar	·027

CROCODILUS HETERODON, Cope.

Alligator heterodon, Proc. American Philosophical Society, 1872, p. 544.

Represented by a single tooth from the posterior part of the mandibular series. It agrees very closely with the corresponding tooth of *Alligator mississippiensis*, but presents some differences. The crown is very low, obtuse, and finely striate; it is compressed and shows a cutting-edge. Its longest diameter is fore and aft; both this and the transverse diameter are proportionately greater than in the modern species. The constriction of the neck is very decided; the fang is large, and especially thick.

Measurements.

	M.
Fore-and-aft diameter of crown	·010
Transverse diameter of crown	·007
Vertical height of crown	

CROCODILUS CLAVIS, Cope.

U. S. Geol. Survey of Terrs. 1872, p. 612.

This species is indicated by the remains of a crocodilian larger than *Crocodilus Elliotii*, but very different from it. The pitting of the dermal scutes, and of all the cranial bones, is deeper than in any other of the Bridger crocodilians in our possession, and very strongly resembles that in *Alligator mississippiensis*.

The *mandible* has a long symphysis, and the alveolar border shows the sudden deep depression just behind it, which is so marked in the ordinary alligator. The rami, however, do not diverge at such an open angle. The jaw is very stout, and is deeper and thinner than in *Crocodilus Elliotii;* and is of about the same proportionate thickness, but deeper than in the alligator. The splenial ends obtusely, and approaches the symphysis somewhat more closely than in the alligator, but does not enter into its formation. The teeth are stout and obtusely conical, not striate, and implanted very close together; they exhibit about the same alternation of size as do those of the alligator. What appears to be the canine is small; it is preceded by a much smaller, and succeeded by a slightly smaller tooth. The dermal scutes are quadrate in shape; they are deeply and irregularly pitted, and have a low but distinct longitudinal keel; they are suturally united by their lateral borders.

CHELONIA.

The Bridger beds have yielded a great abundance and variety of land, marsh, and fresh-water chelonians. They present much variation of form and size; and, owing to the great pressure in the strata, scarcely any two specimens of the same species are alike; occasionally, however, an undistorted specimen is obtained.

The most abundant remains of turtles are those of a species of the recent genus *Emys,* which are found in all the beds, and are, perhaps, the commonest fossils of the basin. The genera *Trionyx* and *Hadrianus* (*Testudo,* Leidy) follow next in order.

HADRIANUS, Cope.

Resembles *Testudo* in form, but has two analscuta, as have most *Emydidæ*.

HADRIANUS ALLABIATUS, Cope.
U. S. Geol. Survey of Territories, 1872, p. 630.

Represented by two ungual phalanges which resemble the claws of *Testudo,* but are proportionately shorter, broader, and not so pointed at the ends. They are oval in section, both longitudinally and transversely; the articular faces are deep and subinferior.

From Henry's Fork.

HADRIANUS OCTONARIUS, Cope.
Geol. Survey of Terrs., 1872. p 630.

Represented by the right humerus of a large individual.

This bone is intermediate in character between the corresponding bones of *Testudo* and *Emys*. The head is longer and narrower than in the former genus; it is implanted on a short, distinct neck, and is strongly convex in both directions. The articular surface extends to the base of the internal tuberosity. The tuberosities, especially the external one, are very large and thick. The external rises above the head, and resembles that of *Emys* in shape. The fossa separating the tuberosities is comparatively narrow. The shaft is much like that of *Emys*, but is somewhat more strongly curved; not so much, however, as in *Testudo*, and the distal end is not so much flattened. The anconeal fossa is wider and deeper than in either genus. The trochlea is broad, and divided by an obscure groove into a small and convex radial face, and a nearly plane (transversely) ulnar face. The condyles are prominent and rugose.

Found near Dry Creek.

EMYS.

EMYS WYOMINGENSIS, Leidy.
Cont. to Ex. Vert. Fauna, p. 140.

Represented by six perfect shells from Cottonwood and Dry Creeks, and Henry's Fork, and a shoulder girdle from Cottonwood Creek, which is provisionally referred to this species.

This species, when adult, was upward of a foot in length, and in the composition of the shell, number of scutes, etc., resembled the living species of the genus.

The Shoulder Girdle has the form characteristic of the *Emydidæ*, but presents some peculiarities. The scapula is short, not being as long as the coracoid, and is not so straight or cylindrical as is usual in *Emys;* but is compressed, and somewhat recurved, and tapers towards the upper extremity, which is marked by a smooth rounded facet for the cartilaginous supra-scapula. The pre-coracoid is long, sub-cylindrical at its origin, but becomes broad and flat distally; at the extremity it is longitudinally striated. The proximal end is flexed forward at an obtuse angle, and is expanded to form the long suture with the scapula and coracoid. It contributes considerably to the glenoid cavity. The coracoid is remarkably long, and after leaving the sutures with the other elements of the girdle, becomes very broad and thin. This is most marked distally. The anterior border is thickened, and the bone is slightly concave on both surfaces.

The three parts of the shoulder girdle are united by distinct bony symphyses; they differ from some of the modern species, in which there is only a ligamentous union between the coracoid and the other parts. It presents a further difference in the long and slender neck which is formed by the flexing of the pre-coracoid, and in the shallow and sub-circular glenoid cavity. The short scapula and long coracoid would indicate a species with a low, broad carapace, as was probably the case in E. *wyomingensis*, though much difficulty has been experienced in determining this point, owing to the distortion of the specimens.

Measurements.

	M.
Length of neck to point between coracoid and precoracoid	.0345
Length of coracoid (about)	.133
Length of precoracoid	.114
Length of scapula (about)	.126

BAENA, Leidy.

BAENA UNDATA, Leidy.
Cont. to Ex. Vert. Fauna, p. 160.

Three nearly perfect shells of different ages, from Dry and Cottonwood Creeks.

TRIONYX, Geoffr.

TRIONYX UINTAENSIS, Leidy.
Cont. to Ex. Vert. Fauna, p. 176.

Cottonwood Creek.

PISCES.

TELEOSTEI.

TELEOCEPHALI.

CYPRINODONTIDÆ.

"Head and body covered with scales; barbels, none. Margin of the upper jaw formed by the premaxillaries only. Teeth in both jaws; upper and lower pharyngeals, with cardiform teeth. Dorsal fin situated on the hinder half of the body." (GÜNTHER.)

TRICOPHANES, Cope.

U. S. Geolog. Survey of the Terrs., 1872, p. 641.

Having secured a very perfect specimen of this genus, we are enabled to complete the definition of its generic characters. We give, then, Professor Cope's definition, with some emendations and additions: Dorsal and anal fins short; ventral fin sometimes beneath and sometimes in advance of the dorsal. The premaxillary forms all of the superior arcade of the mouth, which has a wide gape, opening back behind the orbit. This bone has a row of long, slender, recurved, and subequal teeth implanted in it. The dentary is stout, and has a few small teeth. The branchiostegal rays are six in number, and are rather wide. The preoperculum is serrate. The operculum is ridged on top. The anterior vertebræ are unmodified, and the centra are not elongate. A strong acute spine supports the dorsal, and a similar one the anal fin. There is a long post-clavicle, which may or may not extend to the base of the ventral parallel with the femur. The femur is long and furcate; the external part straight and reaching to the clavicle; the internal curves to meet the corresponding portion of its fellow. The ventral radii are eight in number, and the caudal fin is furčate. The scales are peculiar, and characteristic of the genus. They are very thin, and have borders fringed with long, close-set, bristle-like processes. This genus includes as yet only three species: *T. hians*, from Osivio, Nev.; and *T. foliarum* and *T. Copei*, from Florissant, Col.

TRICOPHANES COPEI, *sp. nov.*

Vertebræ D. 12, C. 18, Radii D. 14, A. 8, V. 8, C. 41, P. 5 only visible, not all preserved. The dorsal fin is long, and projects beyond the beginning of the anal fin; the mouth is terminal, and the muzzle rather sharp. All the bones of the skeleton are very slender; and this is especially true of the ribs, which are not as thick as the interneurals. The neural and hæmal spines are short and slender. The origin of the dorsal is nearer to the muzzle than to the caudal; though, as a whole, the fin is on the hinder half of the body. There are no interneural spines in front of the dorsal fin; those supporting the fin are short, slender, and without the laminar expansions found in *T. foliarum*. The caudal fin consists of numerous jointed rays, which are supported by the neural and hæmal spines of the last three vertebræ. The scales are small, and exhibit the peculiarities of the genus, but the rows are too imperfect to be counted.

Measurements.

	M.
Total length (straight)	·0935
Length of head	·018
Length of vertebral column	·055
Length of caudal fin	·024
Length of dorsal spine	·011
Length of anal spine	·009
Depth of head posteriorly	·016
Length of mandibular ramus	·010
Length to dorsal fin (from muzzle)	·033
Length of dorsal fin	·0135
Depth of body at middle of dorsal fin	·017

Found in the insect beds at Florissant, Col.

CATOSTOMIDÆ.

Body covered with scales; head naked; margin of upper jaw formed by the premaxillaries; mouth toothless. Pharyngeal teeth in a single series, and exceedingly numerous and closely set. Dorsal fin elongate and opposite the ventrals. Anal short, or of moderate length.

AMYZON, Cope.
U. S. Geolog. Survey of the Terrs., 1872, p. 642.

Allied to *Bubalichthys*. Dorsal fin elongate, with a few fulcral spines in front. There are three broad branchiostegals. The vertebræ are short, and the hæmal spines of the caudal fin are distinct and rather narrow. The teeth are arranged comb-like, are truncate, and number from thirty to forty. The dentary bone is slender and toothless, and the angular is distinct.

AMYZON COMMUNE, Cope.
U. S. Geolog. Survey of Terrs., 1873, p.

Represented by several fine specimens from near Castello's Ranch, Col.

SAURODONTIDÆ.

For a very complete definition of this family, see Cope's Cretac. Vert., p. 183.

PORTHEUS, Cope.
Cretac. Vert., p. 189 *et seq.*

"Teeth subcylindric, without serrate or cutting edges, occupying the maxillary, premaxillary, and dentary bones: size, irregular; premaxillary, median maxillary, and dentary much reduced. No foramina on inner face of jaws. Teeth on the premaxillary reduced in number. Opercular and preopercular bones very thin. Cranial bones not sculptured."

PORTHEUS THAUMAS (?), Cope.
Loc. cit. p. 196.

Maxillary large, teeth three; third mandibular small, without cross groove in front of it.

A single tooth from the cretaceous sandstone of Cement Gulch, Col., is provisionally referred to this species; the reference cannot be certain, as the specimens described and figured by Professor Cope have the crowns of the teeth broken. The tooth is not so long as the largest tooth of *P. molossus*, but is of proportionately greater diameter and not so straight. It is conical and recurved, and shows two very obscure cutting-edges, which divide the crown into two unequal faces. It has faint longitudinal striations on all sides, and the apex is sharp.

Measurements.

	M.
Fore-and-aft diameter at the base	0·013
Height of crown	0·020

NEMATOGNATHI.

SILURIDÆ.

"Skin naked, or with osseous scutes, but without scales. Barbels always present; maxillary bone rudimentary, margin of the upper jaw formed by the premaxillaries only. Suboperculum absent; adipose fin present or absent." (Günther.)

RHINEASTES, Cope.

U. S. Geol. Survey of the Terrs., 1872. p. 638.

A genus allied to the recent *Ichthælurus* but differing in the inferior grooving of the vertebræ and in the rough exostoses of the cranial bones. It differs from *Pharcodon* in having the usual band of bristle-like teeth on the dentary.

RHINEASTES ——?

Represented by a portion of the dentary and hyomandibular bones of a large cat-fish from Bridger Butte.

The dentary is broad and shallow, grooved below, and deeply striated on the external face. As much of it as is preserved is straight. The teeth are small, subequal throughout, and very numerous. The hyomandibular is broad and thick, and has a striated surface. The condyle is peculiar in having two separate articular faces above and below, one convex and the other flat.

GANOIDEI.
CYCLOGANOIDEI.
AMIIDÆ.

"Scales cycloid ; a long, soft dorsal fin. Abdominal and caudal parts of the vertebral column subequal in extent." (Günther.)

AMIA.

Body elongate subcylindrical, compressed behind ; snout short and rounded. Jaws with an outer series of closely-set pointed teeth, and with a band of small teeth, similar teeth on the vomer, palatine, and pterygoid bones. Long dorsal, short anal, and rounded nonfurcate caudal fin. Ventrals well developed. A single large gular plate ; branchiostegal rays ten to twelve.

AMIA (Protamia) UINTAENSIS, Leidy.
Cont. to Ext. Vert. Fauna, p 185.

A species of large mud-fishes related to the modern *Amia calva*. The vertebræ are all much wider than they are high. The articular cones have their bottoms considerably above the centre, and are minutely perforate for the notochord. The centrum is transversely curved from side to side, and has the convexity directed forwards ; it is truncate below, making the infero-lateral angles quite prominent in the anterior dorsal region ; in the posterior there are two fossæ. The diapophyses are large, but almost sessile, and take their origin above the centre, on the same line as the bottom of the articular cone. The facets for the neurapophyses are in the form of the figure 8 ; their internal borders are prominent. The atlas has a broad oval centrum, which is not truncate below, and has no markings of any kind on the under surface. The articular faces for the neurapophyses are prominent, and approach near together. The depression for the occipital condyle is small, circular, and situated above the centre.

From Henry's Fork.

AMIA DEPRESSA (?), Marsh.
Pr. Ac. Nat. Sc. 1871, p. 103

In this species the dorsal centra are wide, low, and short,

and of a regular oval outline. The articular cones are shallow, and have their bottoms but slightly above the centre. The neurapophysial facets are deep, with prominent borders, and the diapophyses are sessile. The chief peculiarity is that the under surface of the centrum has no markings of any kind.

Measurements.

	M.
Length of centrum	.010
Breadth of centrum	.040
Depth of centrum	.027

As Professor Marsh has given no measurements, the reference to *A. depressa* cannot be certain.

The specimen was found at Henry's Fork.

AMIA SCUTATA, Cope.
Bull. of U. S. Geol. Survey, No. 1 of Series II, p

A species of about the size of *A. calva*, but with a proportionately larger head. It is represented in our collection by a specimen which lacks only a portion of the caudal and pectoral fins. The *premaxillary* is short and stout, articulates closely with the maxillary, and bears a single row of pointed recurved teeth. These are shorter and straighter than in *A. calva*. The *maxillary* is much as in the modern species, but is rounder. The cranio-facial axis is straight and broad; the *basioccipital* is the largest of the bones; the *basisphenoid* and *presphenoid* are of the usual shape and size, but there is a constriction at their junction which is not marked in the modern species. The *vomer* is long and double, and at the extremity is armed with rows of small teeth. The two portions diverge more perceptibly than in *A. calva*, and are stouter, though this appearance may, to some extent, be due to flattening.

The *frontal* is long, broad, and thick; the upper surface is delicately sculptured in a somewhat different pattern from that seen in the modern species. The *parietal* is short and broad, while the *epiotic* is unusually narrow. The *mandible* is long and slender; the rami are incurved anteriorly, but apparently not to the same extent as in *A. calva*. The articular has the shape of a long and narrow wedge ; it does not form a very close articulation with the dentary. The dentary is long and comparatively slender, has a rounded outline, and is occupied by an external row of large, and an internal band of small teeth. The mandibular teeth, like those of the upper jaw, are somewhat different from the teeth of *A. calva;* they are shorter, stouter, and are not so much incurved or recurved. They are very sharp, and show a constriction below the apex. The small teeth are of the usual size and shape. The jugular plate is well developed, and is long and narrow. The *hyoid arch* is very much the same as that shown in the recent species ; the characteristic flat branchiostegals are well marked ; they appear to be thirteen in number.

The *scapular arch* is long and stout. The *clavicle* is strongly bent, the *supra-clavicle* is short, and the *post-clavicle* is long. The pectoral fin is too indistinct for description.

The *vertebræ* are of considerable depth in the anterior dorsal region, and decrease steadily in size as they go backwards. The neural spines are long and slender, and project strongly backwards. The relation of the centra to the arches seems to be about that seen in *A. calva*, but the neurapophyses are more slender. The diapophyses are long and slender, in this respect differing from the Bridger species and approaching the modern one. The dorsals are thirty-five in number. The caudals number about forty-seven. They have smaller centra, but longer and stronger neurapophyses than the dorsals. The hæmal arch is long, and the hæmapophyses articulate movably with the centra. The spines supporting the caudal fin rays are very stout. The dorsal fin is long and soft, and has long interneurals supporting short rays; these are fifty-three in number. The anal fin, on the other hand, is very short, having but nine rays, which are long and jointed, and articulate with short interhæmals. The caudal fin appears to be of the usual form. The *femur* is of the general shape characteristic of *Amia*, but is not just like that of *A. calva*. It is shorter, broader at the proximal end, while the distal end is narrower, and has a deeper constriction just above it. The ventral fins are placed under the middle of the dorsal region, and have each ten rays. The scales are of the usual cycloidal shape, and minutely striate.

Measurements.

	M.
Total length (estimated in part)	.404
Length of head	.093
Length of vertebral column	.253
Length of caudal fin (estimated)	.058
Length of mandible	.059
Length of jugular plate	.032
Length of dorsal region	.141
Length of caudal region	.112
Depth of body at origin of dorsal fin	
Length of dorsal fin	.117
Length of femur	.027

From the insect beds near Florissant, Col.

PAPPICHTHYS, Cope.
U. S. Geol. Survey of the Terrs., 1872, p. 634.

Vertebræ short, dorsal with projecting diapophyses. Maxillary and dentary bones support but one series of teeth.

PAPPICHTHYS PLICATUS, Cope.
Loc. cit., p. 635.

From Henry's Fork.

PAPPICHTHYS LÆVIS, Cope.
Loc. cit., p. 366.

Represented by three vertebræ from the posterior part of the dorsal region. The centra are subcircular and have deep articular cones, with their bottoms above the centre, and minutely perforate for the notochord. The centra are comparatively quite long, and have prominent projecting rims, and are truncate beneath. The facets for the neurapophyses are long, and are separated into two parts for the contiguous arches. The diapophyses are short and stout.

Measurements.

	M.
Length of centrum	.015
Breadth of centrum	.0315
Depth of centrum	.027

Found at Henry's Fork.

PAPPICHTHYS SYMPHYSIS, Cope.
Loc. cit., p. 636.

Found at Cottonwood Creek.

ELASMOBRANCHII.

PLAGIOSTOMATA.

CARCHARIDAE.

"Eye with a nictitating membrane. An anal fin; two dorsals."

GALEOCERDO.

Caudal fin with a double notch; mouth crescent shaped; teeth subequal in both jaws, oblique, serrate on both margins.

GALEOCERDO FALCATUS, Ag.
Corax falcatus, Poiss. Foss., t. iii., p. 226.

This species is represented by a single tooth from the cretaceous sandstone near the Garden of the Gods, Col. It does not agree exactly with any descriptions or plates which we have seen; but the differences are probably owing to the position or age of the specimen. The crown is low, broad, and not very acute, the edge is finely serrate, and there is no sensible difference between the serrations of the two borders. The anterior border is undulating in outline, but for the most part is convex; the posterior is short and straight, and sends out quite a long heel. One face of the crown is flat and the other is strongly convex. The root is stout and divided into two fangs, which are longer and more distinct than is common in this species. There are no denticles; cementum smooth.

Measurements.

	M.
Length of crown	.017
Depth of fang	.009
Height of crown	.011

GALEOCARDO HARTWELLII, Cope.
Cret. Vert, p., 244.

This species belongs to the group *G. Egertonii*, Ag., having the two edges subequal and symmetrical. A single tooth from Cement Gulch, Col.

CESTRACIONTIDÆ.

"Two dorsal fins, an anal; nasal and buccal cavities confluent. Teeth obtuse, several series being in function." (Günther.)

PTYCHODUS, Ag.
Poissons Fossiles, t. iii., p. 56.

PTYCHODUS WHIPPLEYI, Marcou.
Geology of North America, 1858, p. 33.

Represented by a tooth from the cretaceous of Cement Gulch, almost identical with the specimen described by Dr. Leidy from the cretaceous of Texas. (Cont. to Ext. Vert. Fauna, p. 300.)

SUMMARY.

MAMMALIA.

PRIMATES	4
CARNIVORA	2
PERISSODACTYLA	16
ARTIODACTYLA (?)	1
AMBLYPODA	3
RODENTIA	1
	—27

AVES.

INCERTÆ SEDIS	4
	— 4

REPTILIA.

CROCODILIA	6
CHELONIA	5
	—11

PISCES.

TELEOCEPHALI	3
NEMATOGNATHI	1
CYCLOGANOIDEI	6
PLAGIOSTOMATA	3
INCERTÆ SEDIS	1
	—14

Total	56

EXPLANATION OF PLATE A.
Figure about one fourth natural size
UINTATHERIUM LEIDIANUM.

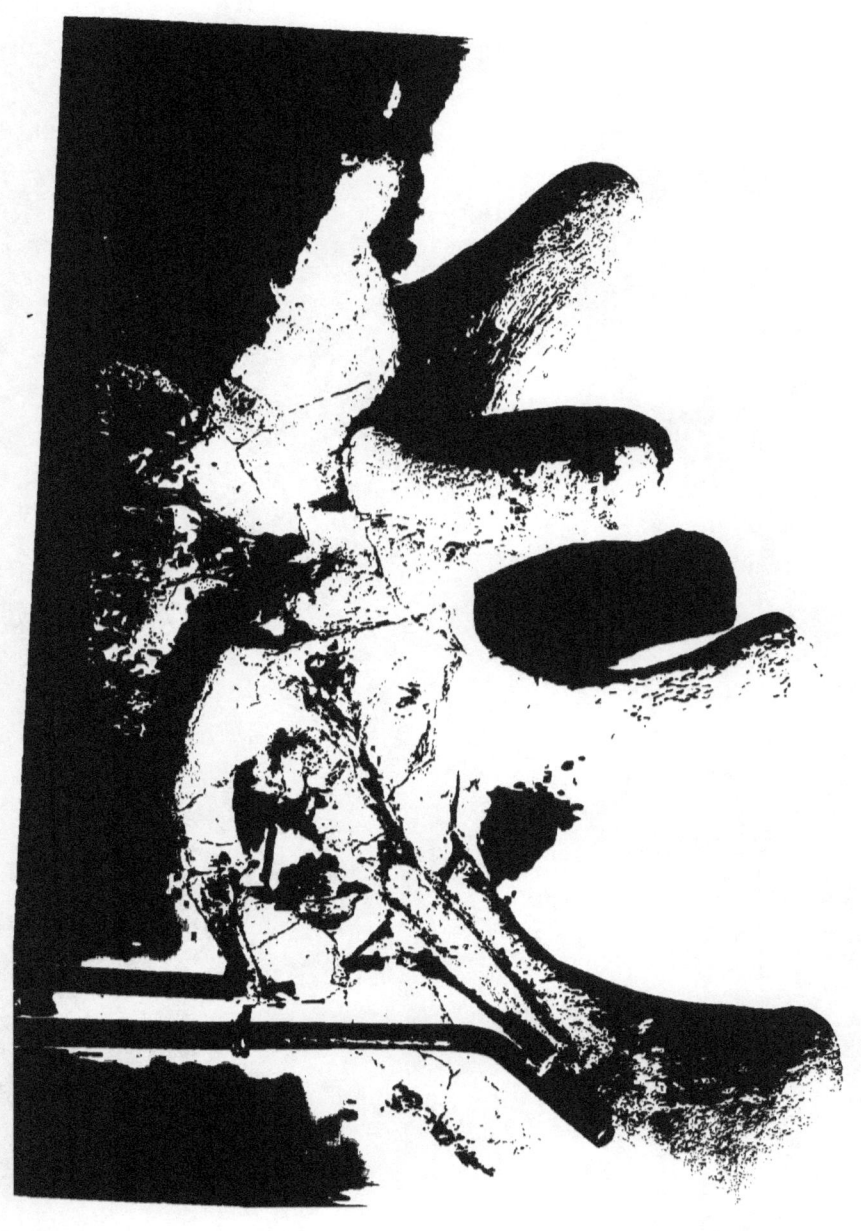

EXPLANATION OF PLATE I.

Figure three fourths natural size.

PALÆOSYOPS PALUDOSUS.

Posterior view of cranium.

PLATE I.

EXPLANATION OF PLATE II.

Figures one fourth natural size.

PALÆOSYOPS MAJOR.

    Fig. 1.—Axis, view of right side.

    Fig. 2.—Axis, anterior view.

    Fig. 3.—Atlas, posterior view.

    Fig. 4.—Atlas, anterior view.

    Figs. 5 and 6.—Seventh cervical anterior and posterior views.

    Figs. 7 and 8.—Anterior and posterior views of right tibia.

    Figs. 9 and 10.—Proximal and distal faces of same.

    Fig. 11.—Proximal end of fibula.

    Fig. 12.—Right femur, posterior view.

    Fig. 13.—Trochlea of the same.

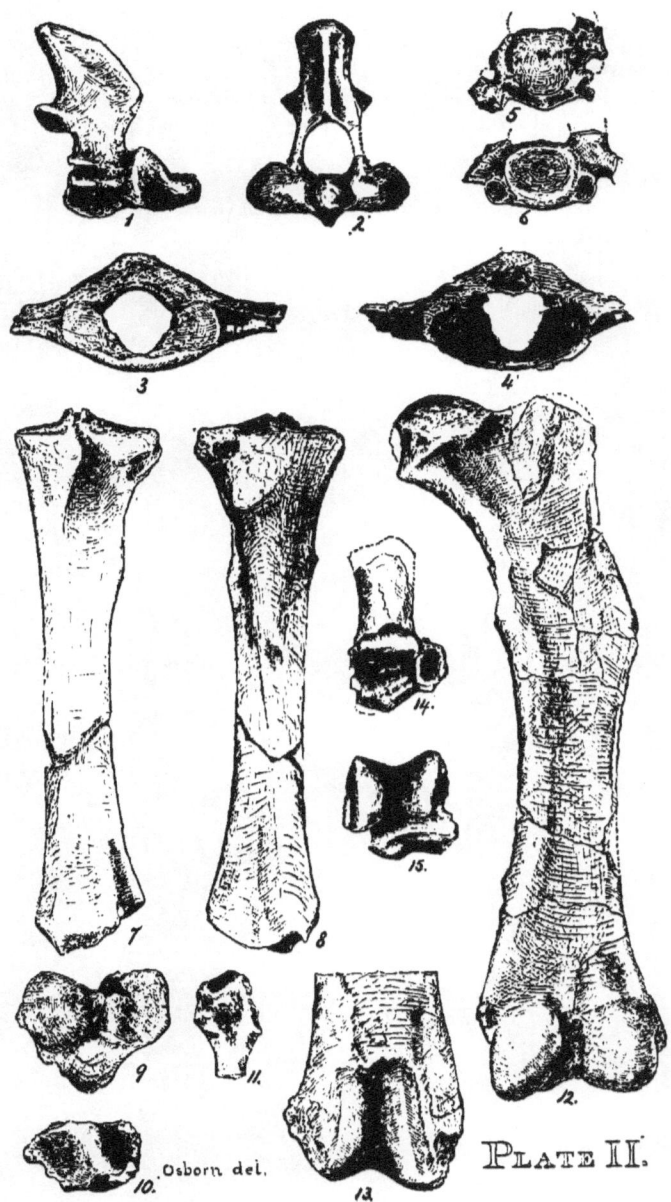

Osborn del.

PLATE II.

EXPLANATION OF PLATE III.
Figures one half natural size.

PALÆOSYOPS PALUDOSUS.

Figs. 1 and 2.—Ulna, anterior view, and distal articular face of same.

Figs. 3, 4, and 5.—Radius, anterior view; and distal and proximal faces of same.

Fig. 6.—Left scapula.

Fig. 7.—Right humerus, proximal end, anterior view.

Fig. 9.—Left manus, anterior view.

PALÆOSYOPS MAJOR.

Fig. 8.—Left humerus, anterior view.

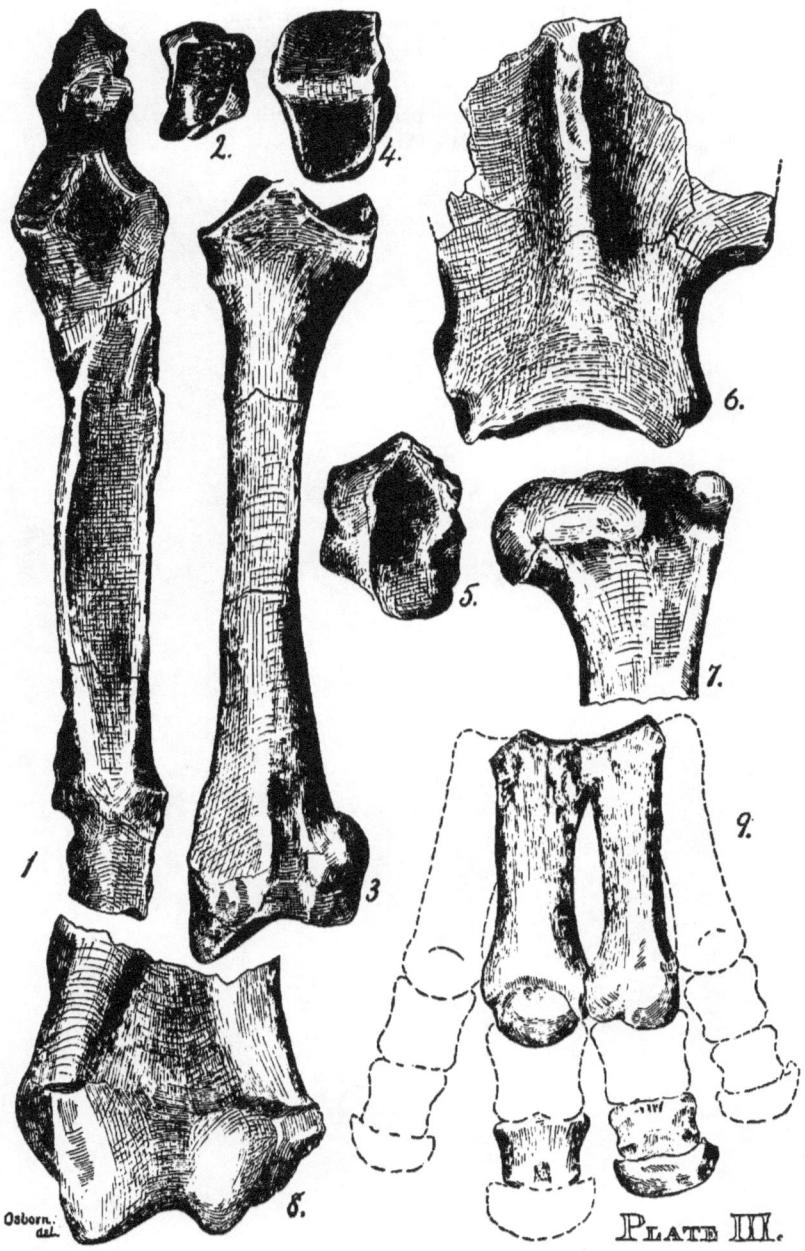

PLATE III.

EXPLANATION OF PLATE IV.

Figure one half natural size.

LEUROCEPHALUS CULTRIDENS.

Upper and lower jaws. View of right side. Dotted outlines indicate probable position of lower incisor and canine series.

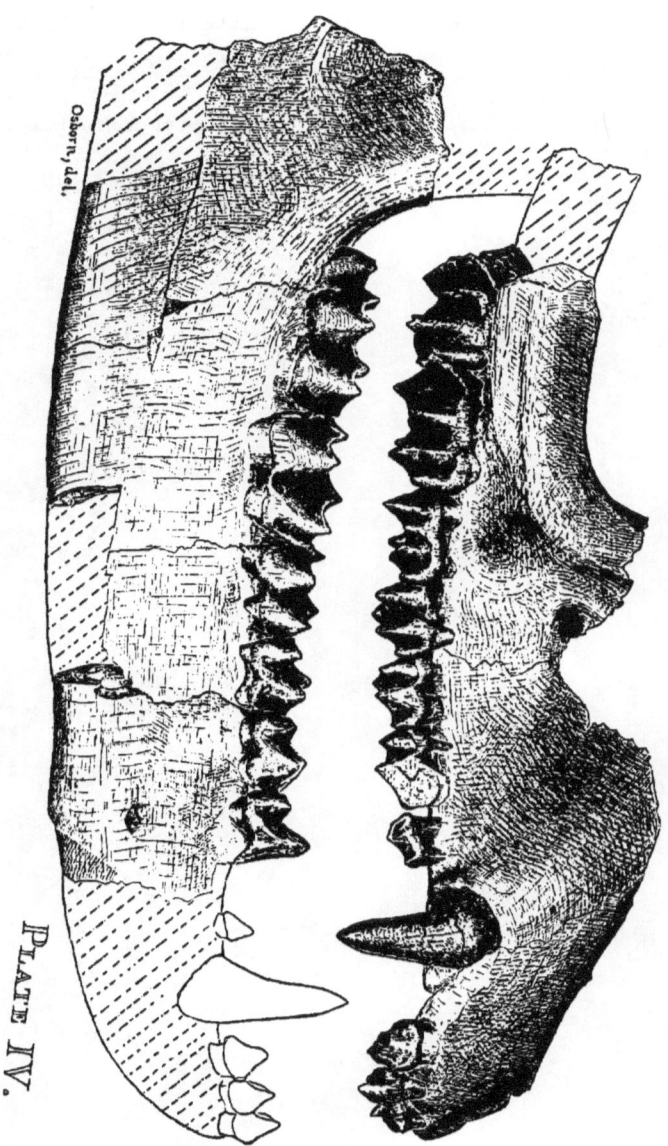

PLATE IV.

EXPLANATION OF PLATE V.

Figure one half natural size.

PALÆOSYOPS MAJOR.

Right innominate bone. The iliac crest and position of pubis estimated in outline.

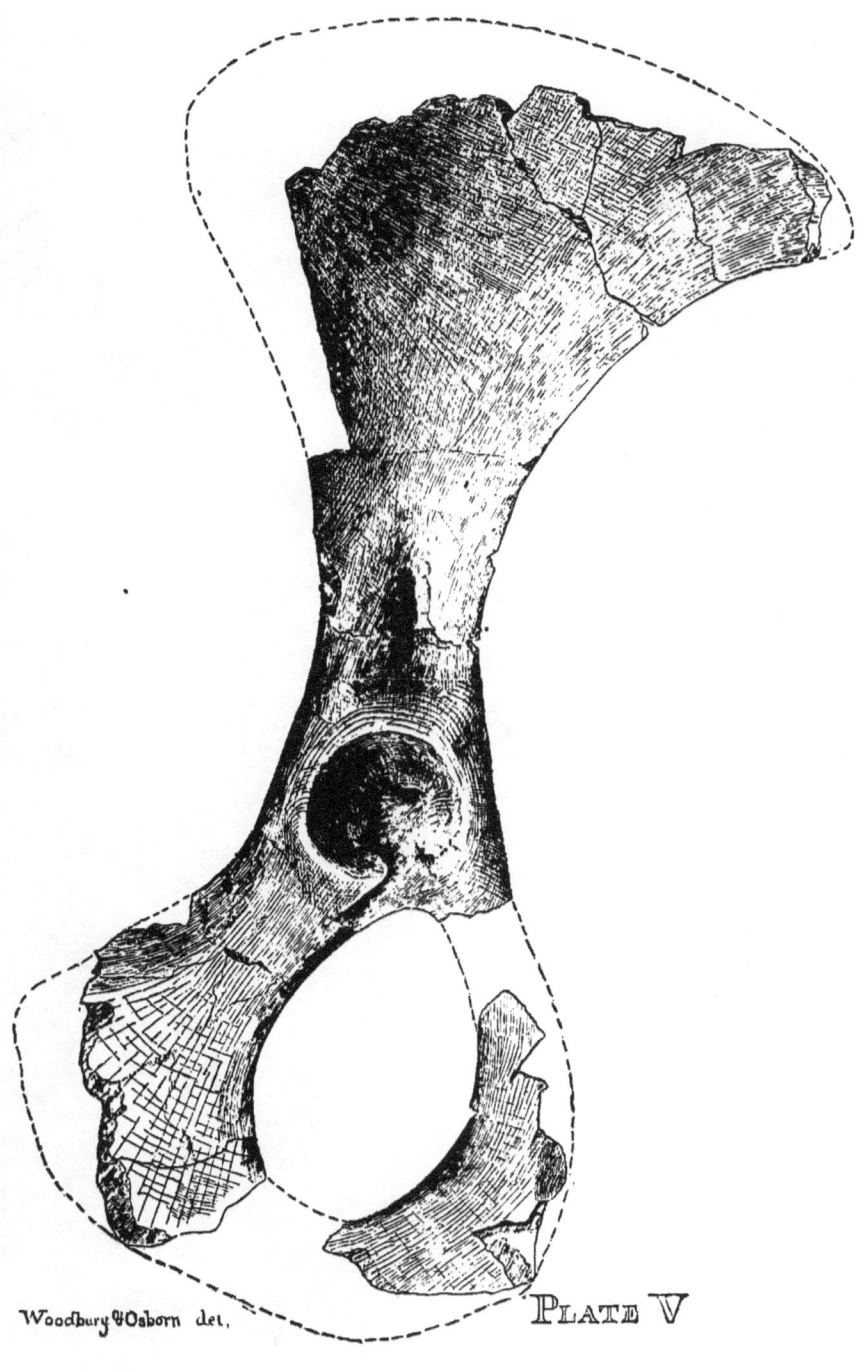

Woodbury & Osborn del.

PLATE V

EXPLANATION OF PLATE VI.

Figures one third natural size.

UINTATHERIUM LEIDIANUM.

    Fig. 1.—Cervical vertebra, fifth or sixth, anterior view.
    Fig. 2.—Dorsal vertebra, anterior region, view of left side.
    Fig. 3.—Dorsal vertebra, middle region, posterior view.
    Fig. 4.—Last lumbar vertebra, view of right side.
    Fig. 5.—Last lumbar vertebra, posterior view.
    Fig. 6.—Caudal series, first four vertebrae.

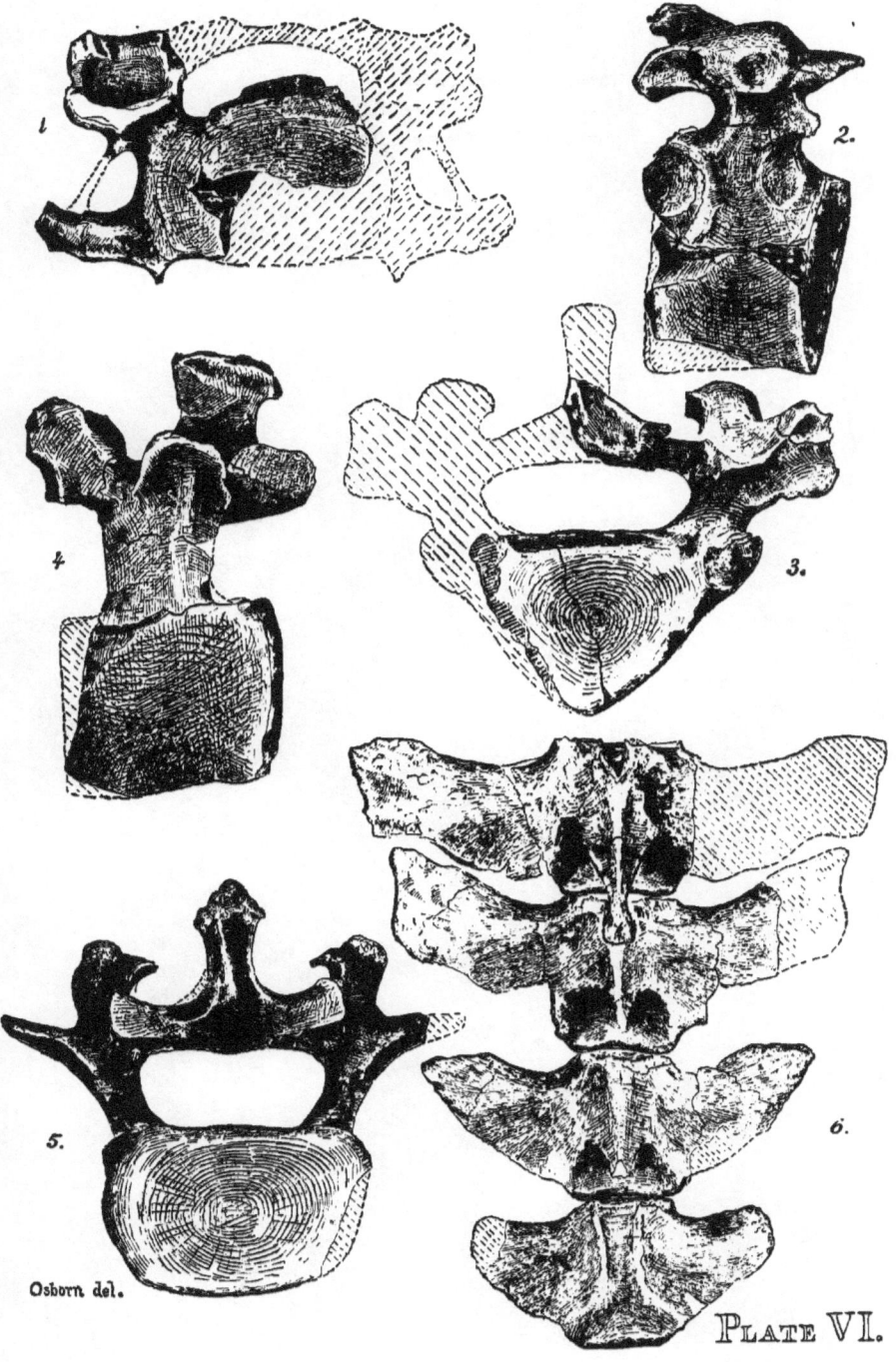

Osborn del.

PLATE VI.

EXPLANATION OF PLATE VII.

Figures one third natural size.

UINTATHERIUM LEIDIANUM.

    Fig. 1.—Anterior view of right humerus.
    Fig. 2.—Ulna, side view, length estimated.

PLATE VII.

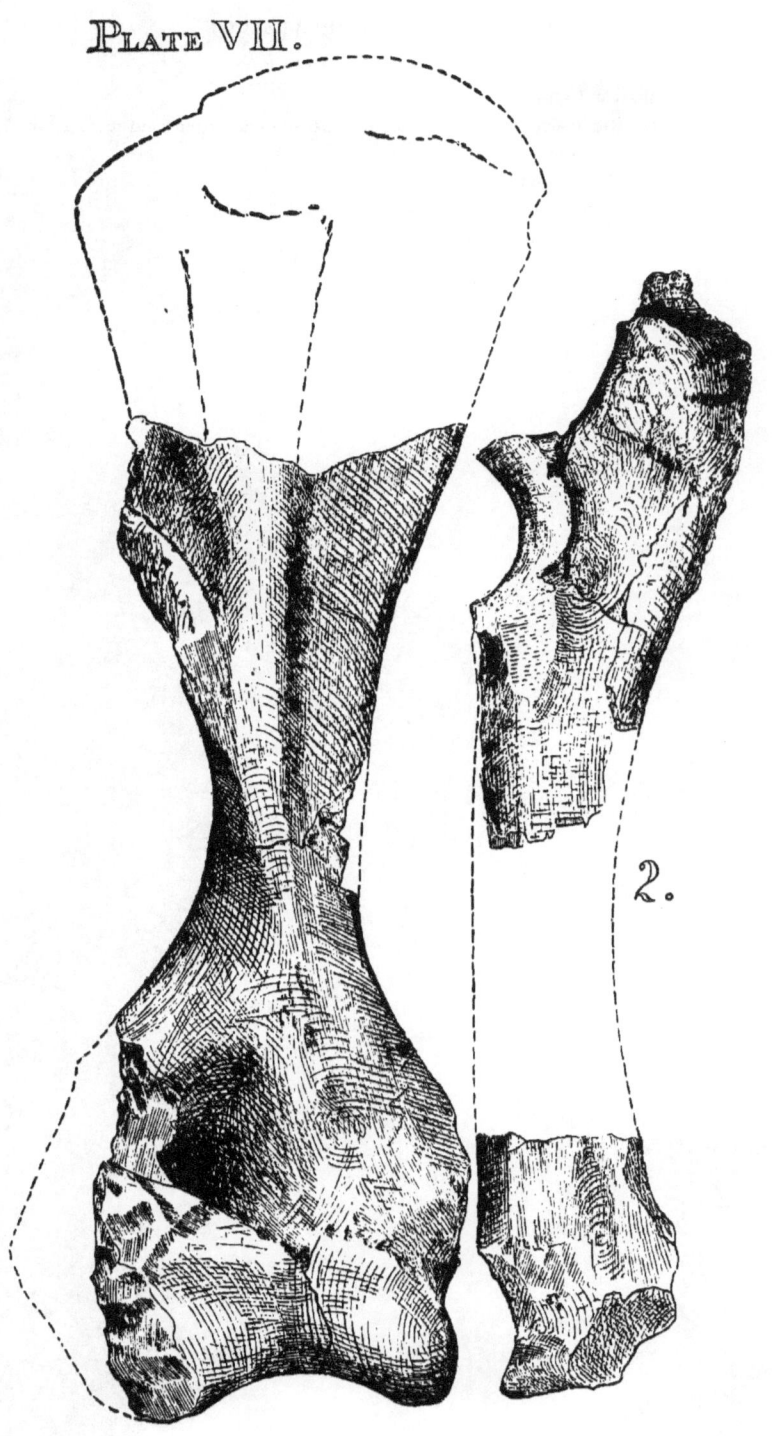

2.

EXPLANATION OF PLATE VIII.

Figures one third natural size.

UINTATHERIUM LEIDIANUM.

Fig. 1.—Right scapula. The dotted outline gives an attempted restoration of the original shape.

Fig. 2.—Tibia of left side, anterior view.

Fig. 3.—View of the proximal articular faces of same.

Fig. 4.—Right femur, posterior view.

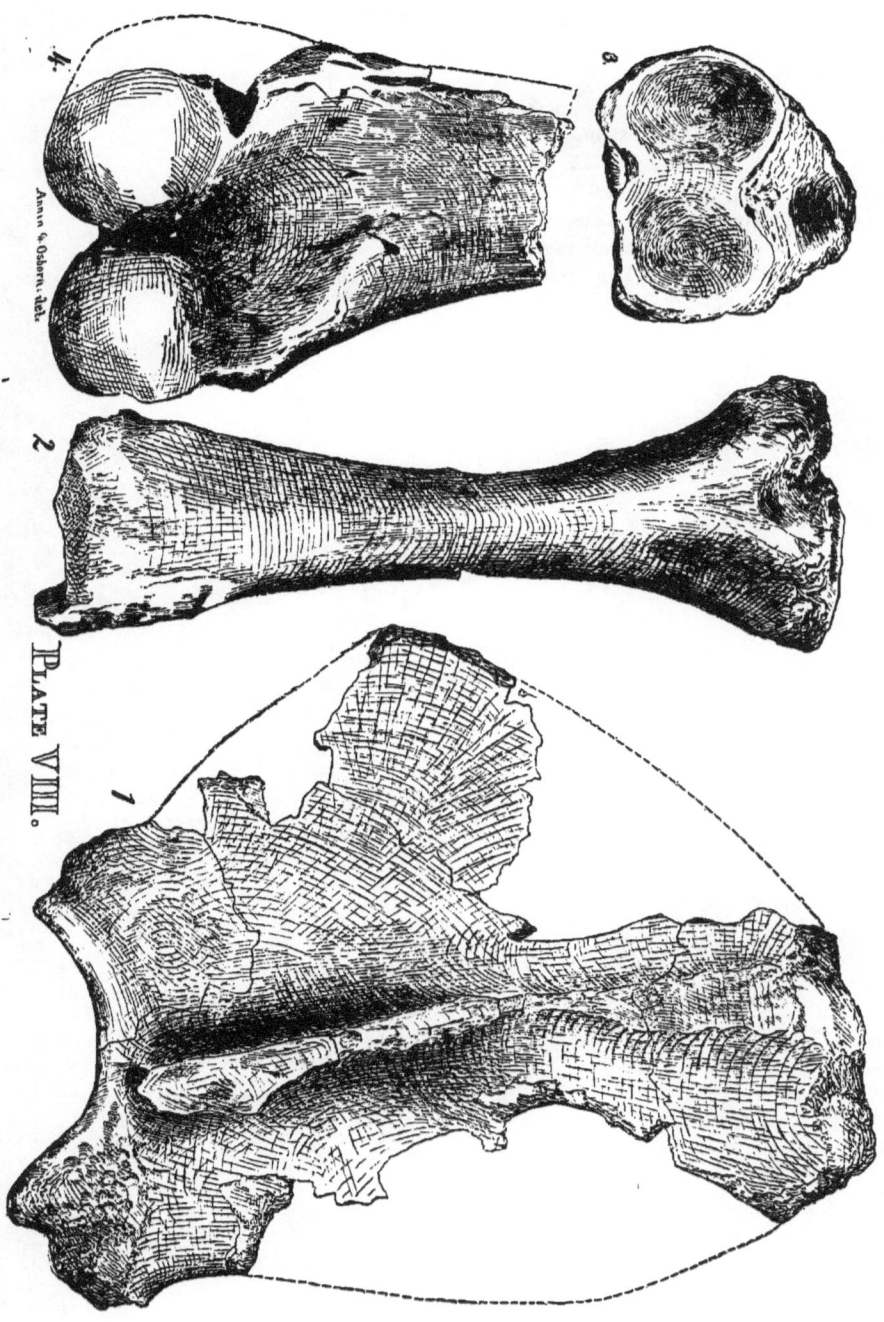

EXPLANATION OF PLATE IX.

Figures four fifths natural size.

OROHIPPUS MAJOR.

Fig. 1.—Right femur, anterior view, length estimated.

Fig. 2.—Right fibula.

Fig. 3.—Right tibia, anterior view.

Fig. 4.—Calcaneum.

Figs. 5 and 6.—Astragalus and navicular.

Fig. 7.—Metatarsals and phalanges.

Last Fig.—A carnivorous sacrum, probably belonging to the brain, described on pp. 20–22.

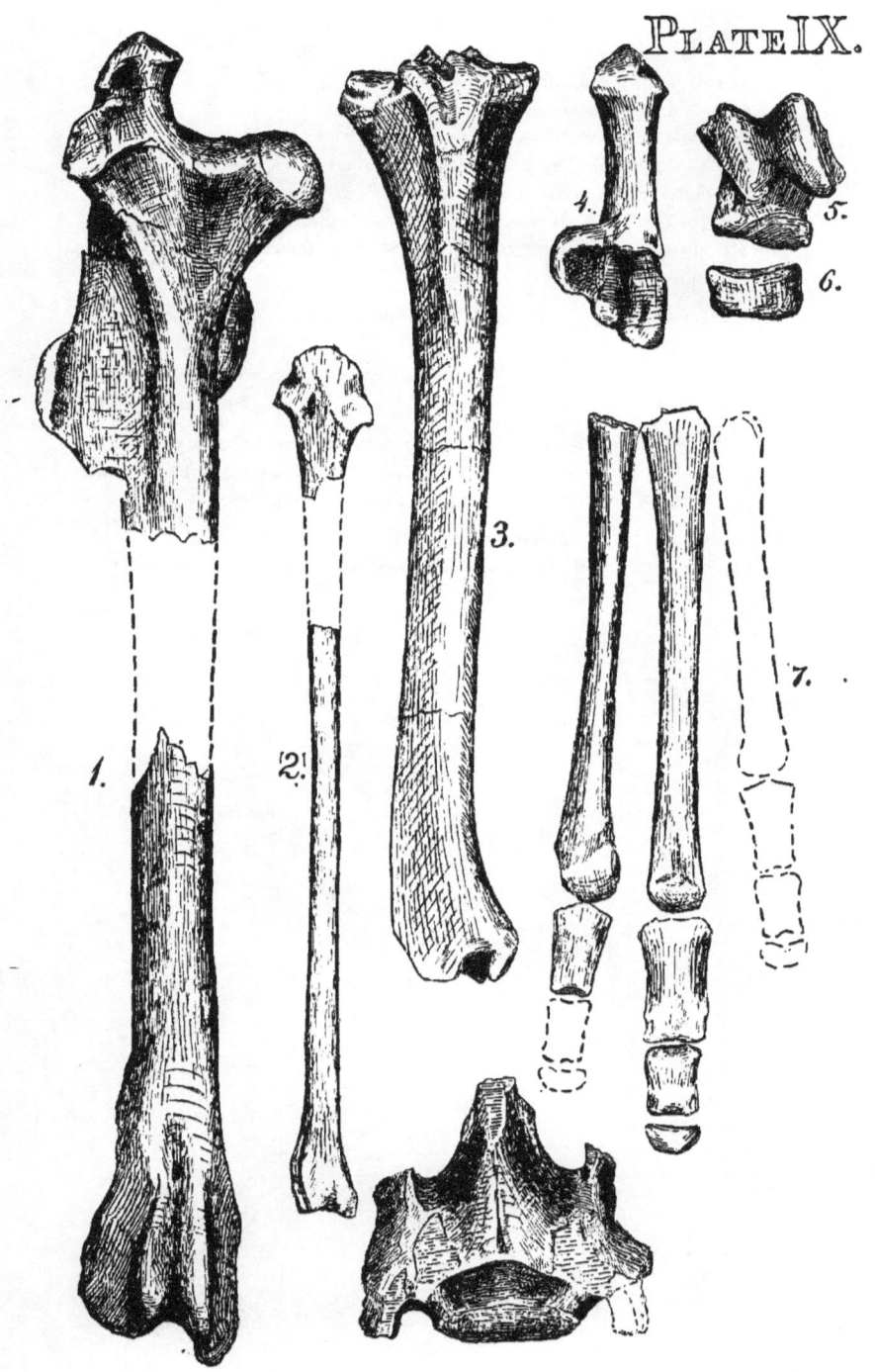

EXPLANATION OF PLATE X.

Figures 1, 2, 3, 4, and 6, natural size.

ITHYGRAMMODON CAMELOIDES.

Fig. 1.—Left premaxillary, inner view.

Fig. 2.—Left premaxillary, outer view.

Fig. 3.—Right maxillary and premaxillary, outer view, premolar, with dotted outline indicating its probable position.

Fig. 4.—The premaxillaries in position (probable), view from above, showing size and position of the incisors. Dotted lines indicate the premaxillary spine (estimated).

PROCAMELUS OCCIDENTALIS, Cope. (See Wheeler's Survey, vol. iv., Plate LXXVII.)

Fig. 5.—Part of right maxillary and premaxillary, showing single incisor with rudimentary alveolus for a second.

PROTOLABIS ———? Cope. (See Wheeler's Survey, vol iv., p. 343.)

Fig. 6.—Right maxillary and premaxillary. (This has never been drawn before, and was kindly lent to us for this purpose by Prof. Cope.)

CAMELUS BACTRIANUS. Modern camel.

Fig. 7.—Right maxillaries, one half natural size.

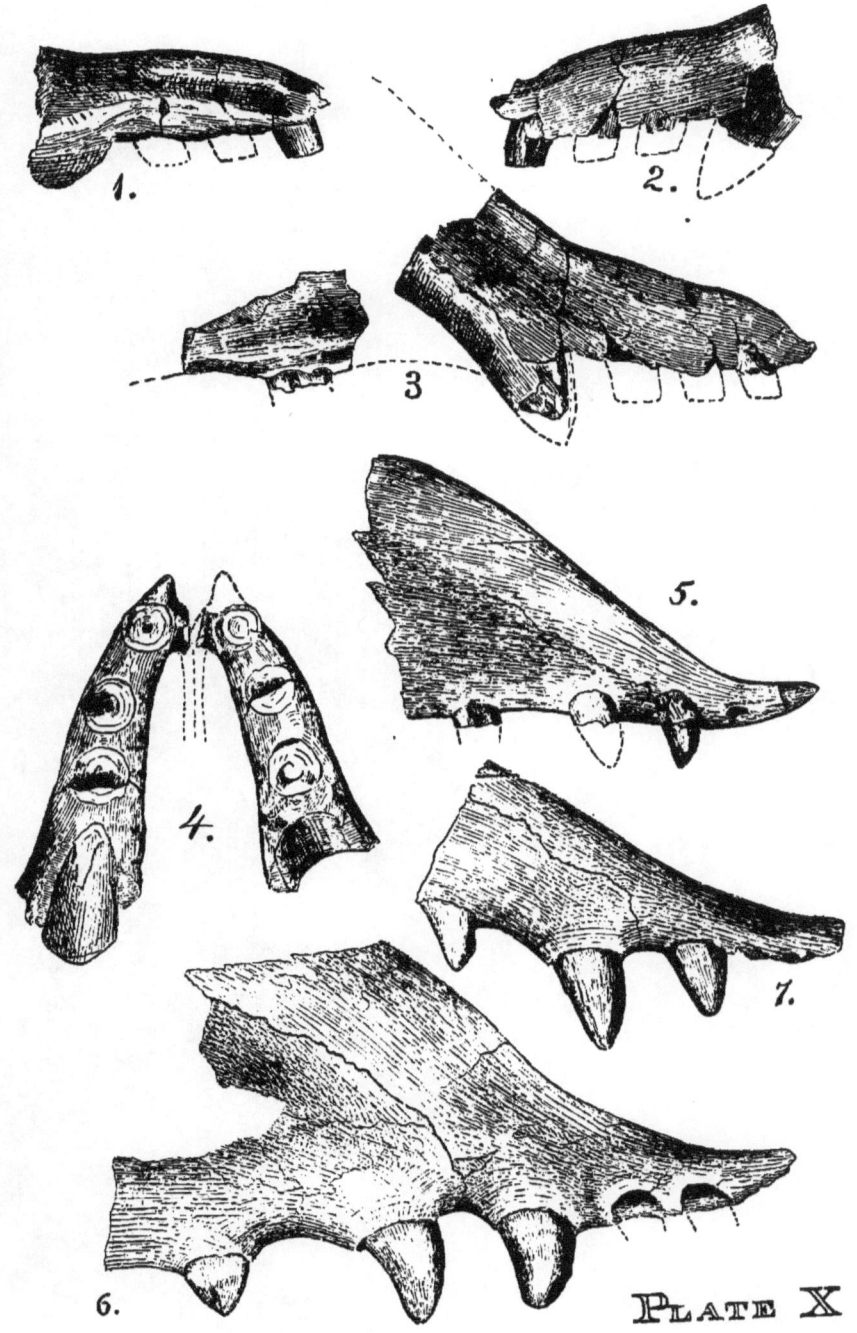

PLATE X

SYSTEMATIC CATALOGUE

OF THE

EOCENE VERTEBRATES OF WYOMING.

ALTHOUGH we have endeavored to make this catalogue as complete and accurate as possible, we feel that errors and omissions are, from the nature of the case, unavoidable. The material is so vast and heterogeneous, and is so much scattered throughout numberless reports, bulletins, journals, and the like, that its correct compilation is attended with great difficulties. However, a beginning is here made which will render subsequent work easier. Owing to the great number of discoveries which have since been made, Dr. Leidy's excellent catalogue of 1871 covers now but a very small portion of the ground.

In this list we have made no attempt to decide disputed questions of priority and synonymy. Doubtless, many of the genera and species here given will be found to be synonyms of American or European forms; but at present these questions cannot be settled. Where synonyms are given, they are the determinations of the original describer indicated in subsequent publications

MAMMALIA.

PRIMATES.

NOTHARCTUS, Leidy. Pr. Ac. Nat. Sc., 1870, p. 114; Cont. to Ext. Vert. Faun., p. 86; Ann. Rep. U. S. Geol. Survey of Terrs., 1871, p. 364.

N. tenebrosus, Leidy, loc. cit.

HYOPSODUS, Leidy. Pr. Ac. Nat. Sc., 1870, p. 110; U. S. Survey of Terrs., 1870, p. 354; do. 1871, p. 362; Cont. to Ext. Vert. Faun., p. 75.
   H. paulus, Leidy, loc. cit.
   H. minusculus, Leidy. Cont. to Ext. Vert. Faun., p. 81.
   H. gracilis, Marsh. Am. Journ. of Sci., vol. ii., p. 42.
   H. vicarius, Cope. Described in U. S. Survey of Terrs., 1872, p. 609, as *Microsyops vicarius*. Wheeler's Survey, vol. iv., pt. ii., p. 150.

MICROSYOPS, Leidy. Pr. Ac. Nat. Sc., 1872, p. 20; U. S. Survey of Terrs., 1871, p. 363; Cont. to Ext. Vert. Faun., p. 82.
   M. gracilis, Leidy, loc. cit. In the "Contributions," Dr. Leidy, considering *Limnotherium*, Marsh, as a synonym of *Microsyops*, has called this species *M. elegans*.

HIPPOSYUS, Leidy. Pr. Ac. Nat. Sc., 1872, p. 37; Cont. to Ext. Vert. Faun., p. 90.
   H. formosus, Leidy, loc. cit.
   H. robustior, Leidy. Cont. to Ext. Vert. Faun., p. 93; *Notharctus robustior*. U. S. Geol. Survey of Terrs., 1871, p. 364.

ANTIACODON, Marsh. Am. Journ. of Sci., 3d Ser. vol. iv., p. 210.
   A. venustus, Marsh, loc. cit.

BATHRODON, Marsh. Am. Journ. of Sci., 3d Ser. vol. iv., p. 211.
   B. typus, Marsh, loc. cit.

LEMURAVUS, Marsh. Am. Journ. of Sci.. 3d Ser. vol. ix., p. 239.
   L. distans, Marsh, loc. cit.

LIMNOTHERIUM, Marsh. Am. Journ. of Sci., 3d Ser. vol. ii., p. 43.
   L. tyrannus, Marsh, loc. cit.
   L. elegans, Marsh, loc. cit.
   L. affine, Marsh. Am Journ. of Sci., 3d Ser. vol. iv., p. 207.

MESACODON, Marsh. Am. Journ. of Sci., 3d Ser. vol. iv., p. 212.
   M. speciosus, Marsh, loc. cit.

PALÆACODON, Leidy. Pr. Ac. Nat. Sc., 1872; p. 21 U. S. Geol. Survey of Terrs., 1871, p. 356; Cont. to Ext. Vert. Faun., p. 122.
   P. verus, Leidy, loc. cit.
   P. vagus, Marsh. Am. Journ. of Sc., 3d Ser. vol. iv., p. 224.

TELMATOLESTES, Marsh. Am. Journ. of Sc., 3d Ser. vol. iv., p. 206.
   T. crassus, Marsh, loc. cit.

TOMITHERIUM, Cope. Pr. Am. Phil. Soc., 1872, p. 470; U. S. Geol. Survey of Terrs., 1872, p. 546; Wheeler's Survey, vol. iv., p. 135, pt. ii.
   T. rostratum, Cope, loc. cit.; U. S. Survey, 1872, p. 548.

ANAPTOMORPHUS, Cope. Pr. Am. Phil. Soc., 1872, p. 554; U. S. Geol. Survey of Terrs., 1872, p. 549.
   A. æmulus, Cope, loc. cit.

OLIGOTOMUS, Cope. On some Eocene mammals, p. 2; U. S. Geol. Survey of Terrs., 1872, p. 607.

   O. cinctus, Cope, loc. cit.

OPISTHOTOMUS, Cope. Wheeler's Survey, vol. iv., p. 151, pt. ii.

   O. astutus, loc. cit., p. 152.

PANTOLESTES, Cope. Pr. Am. Phil. Soc., 1872, p. 467; Wheeler's Survey, v. iv., pt. ii., p. 145.

   P. longicaudus. Pr. Am. Phil. Soc., 1872, p. 467; *Notharctus longicaudus*, U. S. Geol. Survey of Terrs., 1872, p. 549.

SARCOLEMUR, Cope. Pr. Ac. Nat. Sc., 1875, p. 256; Wheeler's Survey, v. iv., pt. ii., p. 147.

   S. pygmæus, Cope. *Lophiotherium pygmæus*, Pr. Am. Phil. Soc., 1872, *extras* July 20; *Antiacodon pygmæus*, U. S. Geol. Survey of Terrs., 1872, p. 607; *Hyopsodus pygmæus*, Pr. Am. Phil. Soc., 1872, p. 461.

   S. furcatus, Cope. *Antiacodon furcatus*, On some Eocene mammals, p. 1.; U. S. Geol. Survey of Terrs., 1872, p. 608.

THINOLESTES, Marsh. Am. Journ. Sc., 3d Ser. vol. iv., p. 205.

   T. anceps, Marsh, loc. cit.

CARNIVORA.

UINTACYON, Leidy. Pr. Ac. Nat. Sc., 1872, p. 277; Cont. to Ext. Vert. Faun., p. 118.

   U. edax, Leidy, loc. cit.
   U. vorax, Leidy, loc. cit. Cont. to Ext. Vert. Faun., p. 120.

SINOPA, Leidy. Pr. Ac. Nat. Sc., 1871, p. 115; U. S. Geol. Survey of Terrs., 1871, p. 355; Cont. to Ext. Vert. Faun., p. 116.

   S. rapax, Leidy, loc. cit.
   S. eximia, Leidy. Cont. to Ext. Vert. Faun., p. 118.

CANIS.

   C. montanus, Marsh. Am. Journ. of Sc., 3d Ser. vol. ii., p. 123.

DROMOCYON, Marsh. Am. Journ. of Sc., 3d Ser. vol. xii., p. 403.

   D. vorax, Marsh, loc. cit.

HARPALODON, Marsh. Am. Journ. of Sc., 3d Ser. vol. iv., p. 216.

   H. sylvestris, Marsh, loc. cit.
   H. vulpinus, Marsh, loc. cit., p. 217.

LIMNOCYON, Marsh. Am. Journ. of Sc., 3d Ser. vol. iv., p. 126.

   L. verus, Marsh, loc. cit.
   L. riparius, Marsh, loc. cit, p. 203.
   L. agilis, Marsh, loc. cit., p. 204.

LIMNOFELIS, Marsh. Am. Journ. of Sc., 3d Ser. vol. iv., p. 202.

   L. ferox, Marsh, loc. cit.
   L. latidens, Marsh, loc. cit., p. 203.

CREOCYON, Marsh. Am. Journ. of Sc., 3d Ser. vol. iv., p. 406
O. latidens, Marsh, loc. cit.

THINOCYON, Marsh. Am. Journ. of Sc., 3d Ser. vol. iv., p. 204.
T. velox, Marsh, loc. cit.

VIVERRAVUS, Marsh. Am. Journ. of Sc., 3d Ser. vol. iv., p. 127.
V. gracilis, Marsh, loc. cit.
V. (?) *nitidus*, Marsh, loc. cit., p. 205.

ZIPHACODON, Marsh. Am. Journ. of Sc., 3d Ser. vol. iv., p. 216.
Z. rugatus, Marsh, loc. cit.

VULPAVUS, Marsh. Am. Journ. of Sc., 3d Ser. vol. ii., p. 124.
V. palustris, Marsh, loc. cit.

MIACIS, Cope. Pr. Am. Phil. Soc., 1872, p. 470.
M. parvivorus, loc. cit.; *Viveravus parvivorus*, U. S. Geol. Survey of Terrs., 1872, p. 560.

MESONYX, Cope. Pr. Am. Phil. Soc., 1872. p. 460; U S. Geol. Survey of Terrs., 1872, p. 550.
M. obtusidens, Cope, loc. cit. U. S. Survey, p. 552.

SYNOPLOTHERIUM, Cope. Pr. Am. Phil. Soc., 1872, p. 483; U. S. Geol. Survey of Terrs., 1872, p. 554.
S. lanius, Cope, loc. cit. U. S. Survey, p. 557.

STYPOLOPHUS, Cope. Pr. Am. Phil. Soc., 1872, p. 466; U. S. Geol. Survey of Terrs., 1872, p. 559; Wheeler's Survey, vol. iv., p. 109.
S. insectivorus, Cope. Pr. Phil. Soc., 1872, p. 469; U. S. Survey, 1872, p. 557.
S. pungens, Cope, loc. cit., pp. 466, 559
S. brevicalcaratus, Cope, loc. cit., pp. 469, 560.

MEGENCEPHALON, gen. nov. This report, p 20.
M. primævus, sp. nov. This report, p. 20.

UNGULATA.
PERISSODACTYLA.

ANCHITHERIUM.
A. (?)—————. This report, p. 23.

OROHIPPUS, Marsh. Am. Journ. of Sc., vol. iv., p. 207; vol. vii., p. 247.
O. pumilus, Marsh, loc. cit., 3d Ser. vol. iv., p. 207.
O. agilis, Marsh, loc. cit., 3d Ser. vol. v., p. 407.
O. major, Marsh, loc. cit., vol. 3d Ser. vii., p. 248.
O. gracilis, Marsh, loc. cit., 3d Ser. vol. vii., p. 249; *Anchitherium gracile*, 3d Ser. vol. ii., p. 38.
O. procyoninus, Cope. Pr. Am. Phil. Soc., 1872, p. 466; U. S. Survey, 1872, p. 606.

PALÆOSYOPS, Leidy. Pr. Ac. Nat. Sc., 1870, p. 113; 1871, pp. 114, 118, 197, 229; 1872, pp. 168, 241; U. S. Geol. Survey, 1870, p. 355; 1871, p. 358; Cont. to Ext. Vert. Faun., p. 27; Cope, U. S. Geol, Survey of Terrs., 1872, p. 591.

P. paludosus, Leidy, loc. cit. Cont. to Ext. Vert. Faun., p. 28.

P. major, Leidy. U. S. Geol. Survey of Montana, 1871, p. 359; Pr. Ac. Nat. Sc., 1872, pp. 168, 241; Cont. to Ext. Vert. Faun., p. 45.

P. humilis, Leidy. Pr. Ac. Nat. Sc., 1872, p. 168, 277; Cont. to Ext. Vert. Faun., p. 58.

P. junius, Leidy. Pr. Ac. Nat. Sc., 1872, p. 277; Cont. to Ext. Vert. Faun., p. 57.

P. lævidens, Cope. Pr. Am. Phil. Soc., 1873; U. S. Survey, 1872, p. 591.

P. vallidens, Cope. Pr. Am. Phil. Soc., 1872, p. 487; loc. cit., p. 592.

P. minor, Marsh. Am. Journ. of Sc., vol. ii., p. 36.

LIMNOHYUS, Leidy. Pr. Ac. Nat. Sc., 1872, p. 242; Cont. to Ext. Vert. Faun., p. 58; Marsh, Am. Journ. of Sc., 1872, 3d Ser. vol. iv., p. 124.

L. laticeps, Marsh. *Palæosyops laticeps*. Am. Journ. Sc. 3d Ser. vol. iv., p. 122.

L. robustus, Marsh, loc. cit., p. 124, is same as *P. major*, Leidy.

L. diaconus, Cope. U. S. Geol. Survey of Terrs., 1872, p. 593.

L. fontinalis, Cope, loc. cit., 594.

TELMATHERIUM, Marsh. Am. Journ. of Sc., 3d Ser. vol. iv., p. 123.

T. validum, Marsh, loc. cit.

LEUROCEPHALUS, gen. nov. This report, p. 42.

L. cultridens, sp. nov. This report, p. 42.

HYRACHYUS, Leidy. Pr. Ac. Nat. Sc., 1871, p. 229; 1872, pp. 19, 163; U. S. Geol. Survey, 1871, p. 360; Cont. to Ext. Vert. Faun., p. 60.

H. agrarius, Leidy, loc. cit.; *H. agrestis*, U. S. Survey, 1871, p. 357.

H. eximius, Leidy. Pr. Ac. Nat. Sc., 1871 p. 229; 1872, p. 163; U. S. Geol. Survey of Terrs., 1871, p. 361; Cont. to Ext. Vert. Faun., p. 66; Cope, U. S. Geol. Survey of Terrs., 1872, p. 595.

H. modestus, Leidy. Pr. Ac. Nat. Sc., 1872, p. 20; U. S. Geol. Survey of Terrs., 1871, p. 361; Cont. to Ext. Vert. Faun., p. 67; *Lophiodon modestus*, Pr. Ac. Nat. Sc., 1870, p. 109.

H. nanus, Leidy. Pr. Ac. Nat. Sc., 3d Ser. 1872, p. 20; U. S. Survey, 1871, p. 361; Cont. to Ext. Vert. Faun., p. 67.

H. princeps, Marsh. Am. Journ. of Sc., 3d Ser. vol. iv., p. 125.

H. implicatus, Cope. On Some Eocene Mammals, p. 5; U. S. Survey, 1872, p 604.

H. crassidens, sp. nov. This report, p. 52.

H. imperialis, sp. nov. This report, p. 50.

H. intermedius, sp. nov. This report, p. 51.

H. paradoxus, sp. nov. This report, p. 53.

HELALETES, Marsh. Am. Journ. of Sc., 3d Ser. vol. iv., p. 218.

H. boöps, Marsh, loc. cit.

H. latidens, sp. nov. This report, p. 54.

LOPHIODON.
   L. bairdianus, Marsh. Am. Journ. of Sc., 3d Ser. vol. ii., p. 36.
   L. affinis, loc. cit., p. 37.
   L. nanus, loc. cit.
   L. pumilus, loc. cit., p. 38.

LOPHIOTHERIUM.
   L. sylvaticum, Leidy. Pr. Ac. Nat. Sc., 1870, p. 126; Cont. to Ext. Vert., Faun, p. 69.
   L. Ballardi, Marsh. Am. Journ. of Sc., 3d Ser. vol. ii., p. 39.

OROTHERIUM, Marsh. Am. Journ. of Sc., 3d Ser. vol. iv., p. 217.
   O. uintanum, Marsh, loc. cit.

HELOHYUS, Marsh. Am. Journ. of Sc., 3d Ser. vol. iv., p. 207.
   H. plicodon, Marsh, loc. cit.

THINOTHERIUM, Marsh. Am. Journ. of Sc., 3d Ser. vol. iv., p. 208.
   T. validum, Marsh, loc. cit.

ARTIODACTYLA.

All the forms described under this head are of uncertain reference.

ELOTHERIUM.
   E. lentum, Marsh. Am. Journ. of Sc., 3d Ser. vol. ii., p. 39.

PLATYGONUS.
   P. Ziegleri, Marsh, loc. cit., p. 40.

PARAHYUS, Marsh. Am. Journ. of Sc., 3d Ser. vol. xii., 402.
   P. vagus, Marsh, loc. cit.

HOMACODON, Marsh. Am. Jour. of Sc., 3d Ser. vol. iv., p. 126.
   H. vagans, Marsh, loc. cit.

ITHYGRAMMODON, gen. nov. This report, p. 56.
   I. cameloides, sp. nov. This report, p. 57.

AMBLYPODA.

Cope. Wheeler's Survey, vol., iv., pt. ii., p. 179.

DINOCERATA.

Marsh. Am. Jour. of Sc., 3d Ser. vol. iv., p. 344; Ibid. vol. v., pp. 117-122, 293; Ibid. vol. vi., p. 300; Ibid. vol. xi., p. 163.

UINTATHERIUM, Leidy. Pr. Ac. Nat. Sc., 1872, p. 169; Am. Journ of Sc., 3d Ser. vol. iv., p. 239; Cont. to Ext. Vert. Faun., p. 96.
   U. robustum, Leidy, loc. cit.; *Uintamastix atrox*, loc. cit.
   U. Leidianum, sp. nov. This report, p. 3.
   U. princeps, sp. nov. This report, p. 81.

TINOCERAS, Marsh. *Titanotherium* (?), Am. Journ. of Sc., 3d Ser. vol. ii., p. 35; *Mastodon*, loc. cit., vol. iv., p. 123, footnote; *Tinoceras*, loc. cit., vol. iv., pp. 322 and 323.
    T. anceps, Marsh, loc. cit., vol. iv., p. 322.
    T. grandis, Marsh, loc. cit., p. 323.

DINOCERAS, Marsh. Am. Jour. of Sc., 3d Ser. vol. iv., p. 344; Ibid. vol. v., pp. 117–122; Ibid. vol. v., p. 408.
    D. mirabilis, Marsh, loc. cit., vol. iv., p. 344.
    D. lucaris, Marsh, loc. cit., vol. v., p. 408.
    D. lacustre, Marsh, loc. cit., vol. iv., p. 344.

LOXOLOPHODON, Cope. Pr. Am. Phil. Soc., 1872, pp. 580, 488; U. S. Survey, 1872, p. 565.
    L. cornutus, Cope, loc. cit.; U. S. Survey, 1872, p. 568.

EOBASILEUS, Cope. Pr. Am. Phil. Soc., 1872, p. 485; U. S. Survey, 1872, p. 575.
    E. pressicornis, Cope. Pr. Am. Phil. Soc., 1872, p. 580; loc. cit.
    E. furcatus, Cope. U. S. Survey, 1872, p. 580; *Loxolophodon furcatus*, Pr. Am. Phil. Soc., 1872, p. 580.

CORYPHODON.
    C. hamatus, Marsh. Am. Journ. of Sc., 3d Ser. vol. xi., p. 425.

BATHMODON, Cope. Pr. Am. Phil. Soc., 1872, p. 417; U. S. Geol. Survey, 1871, p. 350; 1872, p. 586.
    B. radians, Cope, loc. cit.
    B. semicinctus, Cope, loc. cit.
    B. latipes, Cope, loc. cit.

METALOPHODON, Cope. Pr. Am. Phil. Soc., 1872, p. 542; U. S. Geol. Survey, 1872, p. 589.
    M. armatus, Cope, loc. cit.

TILLODONTIA.

Marsh. Am. Jour. of Sc., 3d Ser. vol. ix., p. 221; vol. xi., p. 249.

ANCHIPPODUS, Leidy. Pr. Ac. Nat. Soc., 1868, p. 232; Ext. Mam. N. Am., p. 403.
    Trogosus. Pr. Ac., 1871, p. 113; Cont. to Ext. Vert. Faun., p. 71.
    A. riparius, Leidy. Pr. Ac. Nat. Soc., 1868, p. 232; Cont. to Ext. Vert. Faun., p. 71 (as *Trogosus castoridens*).
    A. vetulus, Leidy; *Trogosus vetulus*, Pr. Ac. Nat. Sc., 1871, p. 229; Cont. to Ext. Vert. Faun., p. 75.

TILLOTHERIUM. Am. Jour. of Sc., 3d Ser. vol. v., p. 485; Ibid. vol. xi., p. 249.
    T. hydracoides, Marsh, loc. cit.
    T. latidens, Marsh, loc. cit., vol. vii., p. 533.
    T. fodiens, Marsh, loc. cit., vol. ix., p. 241.

STYLINODON, Marsh. Am. Jour. of Sc., 3d Ser. vol. vii., p. 532.
    S. mirus, loc. cit.

RODENTIA.

PARAMYS, Leidy. Pr. Ac. Nat. Sc., 1871, p. 231; U. S. Geol. Survey, 1871, p. 357; Cont. to Ext. Vert. Faun., p. 110.
   P. delicatus, Leidy, loc. cit.
   P. delicatior, Leidy, loc. cit.
   P. delicatissimus, Leidy, loc. cit.; Cont. to Ext. Vert. Faun., p. 111.
   P. robustus, Marsh. Am. Jour. of Sc., 3d Ser. vol. iv., p. 218.
   P. superbus, sp. nov. This report, p. 84.
   P. leptodus, Cope. U. S. Geol. Survey, 1872, p. 609.

MYSOPS, Leidy. Pr. Ac. Nat. Sc., 1871, p. 232; U. S. Geol. Survey, 1871, p. 357; Cont. to Ext. Vert. Faun., p. 111.
   M. minimus, Leidy, loc. cit.
   M. fraternus, Leidy. Cont. to Ext. Vert. Faun., p. 112.

PSEUDOTOMUS, Cope. Pr. Am. Phil. Soc., 1872, p. 467; U. S. Geol. Survey, 1872, p. 610.
   P. hians, Cope, loc. cit.

ARCTOMYS, Marsh. Am. Jour. of Sc., 3d Ser. vol. ii., p. 121.
   A. vetus, Marsh, loc. cit.

GEOMYS, Marsh. Am. Jour. of Sc., 3d Ser. vol. ii., p. 121.
   G. bisulcatus, Marsh, loc. cit.

SCIURAVUS, Marsh. Am. Jour. of Sc., 3d Ser. vol. ii., p. 122.
   S. nitidus, Marsh, loc. cit.
   S. undans, Marsh, loc. cit.
   S. parvidens, Marsh, loc. cit., 3d Ser. vol. iv., p. 220.

TILLOMYS, Marsh. Am. Jour. of Sc., 3d Ser. vol. iv., p. 219.
   T. senex, Marsh, loc. cit.
   T. parvus, Marsh, loc. cit.

TACHYMYS, Marsh. Am. Jour. of Sc., 3d Ser. vol. iv., p. 219.
   T. lucaris, Marsh, loc. cit.

COLONYMYS, Marsh. Am. Jour. of Sc., 3d Ser. vol. iv., p. 220.
   C. celer, Marsh, loc. cit.

INSECTIVORA.

OMOMYS, Leidy. Pr. Ac. Nat. Sc., 1869, p. 63; Ext. Mam. of N. Am., p. 408; Cont. to Ext. Vert. Faun., p. 120.
   O. Carteri, Leidy, loc. cit.

WASHAKIUS, Leidy. Cont. to. Ext. Vert. Faun., p. 123.
   W. insignis, Leidy, loc. cit.

PASSALACODON, Marsh. Am. Jour. of Sc., vol. iv. p. 208.
   P. litoralis, Marsh, loc. cit.

ANISACODON,* Marsh. Am. Journ. of Sc., 3d Ser. vol. iv., p. 209.
    A. elegans, Marsh, loc. cit.

CENTETODON, Marsh, Am. Journ. of Sc., 3d Ser. vol. iv., p. 209.
    C. pulcher, Marsh, loc. cit.
    C. altidens, Marsh, loc. cit., p. 214.

HEMIACODON, Marsh, Am. Journ. of Sc., 3d Ser. vol. iv., p. 212.
    H. gracilis, Marsh, loc. cit.
    H. nanus, Marsh, loc. cit., p. 113
    H. pucillus, Marsh, loc. cit.

ENTOMODON, Marsh, Am. Journ. of Sc., 3d Ser. vol. iv., p. 214.
    E. comptus, Marsh, loc. cit.

ENTOMACODON, Marsh, Am. Journ. of Sc., 3d Ser. vol. iv., p. 214.
    E. minutus, Marsh, loc. cit.
    E. angustidens, Marsh, loc. cit., p. 222.

APATEMYS, Marsh, Am. Journ. of Sc., 3d Ser. vol. iv., p. 221.
    A. bellus, Marsh, loc. cit.
    A. bellulus, Marsh, loc. cit.

TALPAVUS, Marsh, Am. Journ. of Sc., 3d Ser. vol. iv., p. 128.
    T. nitidus, Marsh, loc. cit.

CHIROPTERA.

NYCTITHERIUM, Marsh. Am. Journ. of Sc., 3d Ser. vol. iv., p. 127.
    N. velox, Marsh, loc. cit.
    N. priscum, Marsh, loc. cit., p. 128.

NYCTILESTES, Marsh. Am. Journ. of Sc., 3d Ser. vol. iv., p. 215.
    N. serotinus, Marsh, loc. cit.

MARSUPIALIA.

TRIACODON, Marsh. Am. Journ. of Sc., 3d Ser. vol. ii., p. 123.
    T. fallax, Marsh, loc. cit.
    T. grandis, Marsh, loc. cit. 3d Ser. vol. iv., p. 222.
    T. nanus, Marsh, loc. cit., p. 223.
    T. aculeatus, Cope. Pr. Phil. Soc., 1872, p. 460; U. S. Geol. of Terrs., 1872, p. 611.

Genera incertæ sedis.

STENACODON, Marsh. Am. Journ. of Sc., 3d Ser. vol. iv., p. 210.
    S. rarus, Marsh, loc. cit.

---

* Professor Marsh has subsequently used this name to designate a genus of the *Brontotheridæ*. Am. Jour. of Sc., vol. ix., p. 246.

AVES.

RAPTORES.

BUBO.
 B. leptosteus, Marsh. Am. Journ. of Sc., 3d Ser. vol. ii, p. 126.

GRALLATORES.

ALETORNIS, Marsh. Am. Journ. of Sc., 3d Ser. vol. iv., p. 256.
 A. nobilis, Marsh, loc. cit.
 A. pernix, Marsh, loc. cit.
 A. venustus, Marsh, loc. cit., p. 257.
 A. gracilis, Marsh, loc. cit., p. 258.
 A. bellus, Marsh, loc. cit.

SCANSORES.

UINTORNIS, Marsh. Am. Journ. of Sc., 3d Ser. vol. iv., p. 259.
 U. lucaris, Marsh, loc. cit.

REPTILIA.

CROCODILIA.

CROCODILUS.
 C. liodon, Marsh. Am. Journ. of Sc., 3d Ser. vol. i., p. 454.
 C. affinis, Marsh, loc. cit.
 C. Grinnellii, Marsh, loc. cit., p. 455.
 C. brevicollis, Marsh, loc. cit., p. 456.
 C. parvus, sp. nov. This report, p. 91.
 C. clavis, Cope. Pr. Am. Phil. Soc., 1872, p. 485; U. S. Geol. Survey, 1872, p. 612.
 C. sulciferus, Cope. Pr. Am. Phil. Soc., 1872, p. 555; U. S. Survey, loc. cit.
 C. heterodon, Cope. *Alligator heterodon*, Pr Am. Phil. Soc., 1872, p. 544; U. S. Geol. Survey, 1872, p. 614.
 C. aptus, Leidy. Cont. to Ext. Vert. Faun., p. 126.
 C. Elliotii, Leidy, loc. cit.

DIPLOCYNODUS.
 D. subulatus, Cope. U. S. Geol. Survey, 1872, p. 613; *Crocodilus subulatus*, Pr. Am. Phil. Soc., 1872, p. 554.
 D. polyodon, Cope. U. S. Geol. Survey, 1872, p. 614. (In his report to Lt. Wheeler, Prof. Cope says that "a single species, the *D. subulatus*, occurs" in the Bridger Basin, (p. 60). It is therefore probable that he has changed the *D. polyodon*, but we cannot find its synonym.)

LACERTILIA.

SANIVA, Leidy. Pr. Ac. Nat. Sc., 1870, p. 124 ; U. S. Survey, 1870, p. 368 ; do., 1871, p. 370 ; Cont. to Ext. Vert. Faun., p. 181.

   S. ensidens, Leidy, loc. cit.
   S. major, Leidy. Cont. to Ext. Vert. Faun., p. 182.

CHAMELEO.

   C. pristinus, Leidy. Pr. Ac. Nat. Sc., 1872, p. 277 ; Cont. to Ext. Vert. Faun., p. 184.

NAOCEPHALUS, Cope. Pr. Am. Phil. Soc., 1872, p. 465 ; U. S. Geol. Survey, 1872, p. 631.

   N. porrectus, Cope, loc. cit.; U. S. Geol. Survey, 1872, p. 632.

GLYPTOSAURUS, Marsh. Pr. Ac. Nat. Sc., 1871, p. 105 ; Am. Journ. of Sc., 3d Ser. vol. i., p. 456.

   G. sylvestris, Marsh, loc. cit.
   G. nodosus, Marsh, Am. Journ. of Sc., vol. i., p. 458.
   G. ocellatus, Marsh, loc. cit., vol. i., p. 458, and vol. iv., p. 306.
   G. anceps, Marsh, loc. cit., vol. i., p. 458.
   G. princeps, Marsh, loc. cit., 3d Ser. vol. iv., p. 302.
   G. brevidens, Marsh, loc. cit., p. 305.
   G. rugosus, Marsh, loc. cit.
   G. sphenodon, Marsh, loc. cit., p. 306.

THINOSAURUS, Marsh. Am. Journ. of Sc., 3 Ser. vol. iv., p. 299.

   T. paucidens, Marsh, loc. cit.
   T. leptodus, Marsh, loc. cit., p. 300.
   T. crassus, Marsh, loc. cit., p. 301.
   T. grandis, Marsh, loc. cit.
   T. agilis, Marsh, loc. cit., p. 302.

OREOSAURUS, Marsh. Am. Journ. of Sc., 3d Ser. vol. iv., p. 303.

   O. vagans, Marsh, loc. cit.
   O. lentus, Marsh, loc. cit., p. 307.
   O. gracilis, Marsh, loc. cit.
   O. microdus, Marsh, loc. cit., p. 308.
   O. minutus, Marsh, loc. cit.

TINOSAURUS, Marsh. Am. Journ. of Sc., 3d Ser. vol. iv., p. 304.

   T. stenodon, Marsh, loc. cit.
   T. lepidus, Marsh, loc. cit., p. 308.

IGUANAVUS, Marsh. Am. Journ. of Sc., 3d Ser. vol, iv., p. 309.

   I. exilis, Marsh, loc. cit.

LIMNOSAURUS, Marsh. Am. Journ. of Sc., 3d Ser.vol. iv., p. 309.

   L. ziphodon, Marsh, loc. cit. ; *Crocodilus ziphodon*, loc. cit., vol. i., p. 453.

OPHIDIA.

BOAVUS, Marsh. Am. Journ. of Sc., 3d Ser. vol. i., p. 323.
  B. occidentalis, Marsh, loc. cit.
  B. agilis, Marsh, loc. cit., p. 324.
  B. brevis, Marsh, loc. cit.

LITHOPHIS, Marsh. Am. Journ. of Sc., 3d Ser. vol. i., p. 325.
  L. Sargenti, Marsh, loc. cit.

LIMNOPHIS, Marsh. Am. Journ. of Sc, 3d Ser. vol. i., p. 326.
  L. crassus, Marsh, loc. cit.

PROTAGRAS, Cope. Pr. Am. Phil. Soc., 1872, p. 471 : U. S. Geol.Survey, 1872, p. 632.
  P. lacustris, Cope, loc. cit.

CHELONIA.

TESTUDO.
  T. Corsoni, Leidy. Pr. Ac. Nat. Sc., 1871, p. 154 ; 1872, p. 268 ; U. S. Geol. Survey, 1871, p. 366 ; Cont. to Ext. Vert. Faun., p. 132 ; Emys Carteri, Pr. Ac. Nat. Sc., 1871, p. 228 ; U. S. Geol. Survey, 1871, p. 367. (Professor Cope refers this species to his Hadrianus.)

HADRIANUS, Cope. Pr. Am. Phil. Soc., 1872, p. 468 ; U. S. Geol. Survey of Terrs.,1872, p. 630.
  H. allabiatus, Cope. Pr. Am. Phil. Soc., 1872, p. 471 ; U. S. Survey, loc. cit.
  H. octonarius, Cope, loc. cit. Pr. Am. Phil. Soc. 1872, p. 468.

EMYS.

  E. wyomingensis, Leidy. Pr. Ac. Nat. Sc., 1869, p. 66 ; U. S. Geol. Survey, 1871, p. 367 ; Cont. to Ext. Vert. Faun., p. 140. E. Jeansii, Pr. Ac. Nat. Sc., 1870, p. 123 ; E. stevensoniansis, loc. cit., p. 5. E. Haydeni, loc. cit., p. 123.
  E. septarius, Cope. U. S. Geol. Survey, 1872, p. 625.
  E. latilabiatus, Cope. Pr. Am. Phil. Soc., 1872, p. 471 ; loc. cit., p. 626.
  E. gravis, Cope. Notomorpha gravis and N. Garmanii, Pr. Am. Phil. Soc., 1872, pp. 476–77 ; E. gravis, U. S. Survey, 1872, p. 626.
  E. testudineus, Cope. U. S. Geol. Survey, 1872, p. 627 ; Notomorpha testudineus, Pr. Am. Phil. Soc., 1872, p. 475.
  E. euthnætus, Cope. U. S. Geol. Survey, 1872, p. 628.
  E. megaulax, Cope, loc. cit.
  E. pachylomus, Cope, loc. cit., p. 629.
  E. terrestris, Cope, loc. cit. ; Palæotheca terrestris, Pr. Am. Phil. Soc., 1872, p. 464.
  E. polycyphus, Cope, loc. cit., 630 ; P. polycyphus, loc. cit., p. 463.

HYBEMYS, Leidy. Pr. Ac. Nat. Sc., 1871 ; p. 103 ; Cont. to Ext. Vert. Faun., p. 174.
  H. arenarius, Leidy, loc. cit.

BAPTEMYS, Leidy. Pr. Ac. Nat. Sc., 1870, p. 4 ; U. S. Survey, 1870, p. 367 ; do., 1872, p. 367 ; Cont. to Ext. Vert. Faun., 157. (Professor Cope refers this genus to *Dermatemys Gray*).

B. wyomingensis, Leidy, loc. cit.

CHISTERNON, Leidy. Pr. Ac. Nat. Sc., 1872, p. 162.

C. undatum. Cont. to Ext. Vert. Faun., p. 341 ; *Baena undata*, Pr. Ac. Nat. Sc., 1871, p. 228 ; U. S. Survey, 1871, p. 369 ; Cont. to Ext. Vert. Faun., p. 161.

BAENA, Leidy. Pr. Ac. Nat. Sc., 1871, p. 228 ; U. S. Survey, 1870, p. 367 ; do., 1871, p. 368 ; Cont. to Ext. Vert. Faun., p. 161.

B. arenosa, Leidy, loc. cit. ; *Baena affinis*, U. S. Survey, 1870, p. 367.

B. hebraica, Cope. Pr. Am. Phil. Soc., 1872, p. 463 ; U. S. Survey, 1872, p. 621.

B. ponderosa, Cope. U. S. Survey, 1872, p. 624.

ANOSTEIRA, Leidy. Pr. Ac. Nat. Sc., 1871, p. 102, 114 ; U. S. Survey, 1871, p. 370 ; Cont. to Ext. Vert. Faun., p. 174.

A. ornata, Leidy, loc. cit.

A. radulina, Cope. Pr. Am. Phil. Soc., 1872, p. 555 ; U. S. Survey, 1872, p. 650.

TRIONYX.

T. guttatus, Leidy. Pr. Ac. Nat. Sc., 1869, p. 66 ; 1870, p. 5 ; 1871, p. 228 ; U. S Geol. Survey, 1870, p. 367 ; do., 1871, p. 370 ; Cont. to Ext. Vert. Faun., p. 176.

T. uintaensis, Leidy. Pr. Ac. Nat. Sc., 1872, p. 267 ; Cont. to Ext. Vert. Faun., p. 178.

T. heteroglyptus, Cope. U. S. Geol, Survey, 1872, p. 616.

T. scutumantiquum, Cope, loc. cit., p. 617.

T. concentricus, Cope, loc. cit.; Pr. Am. Phil. Soc., 1872, p. 469.

AXESTUS, Cope. Pr. Am. Phil. Sc., 1872, p. 462 ; U. S. Geol. Survey, 1872, p. 615.

A. byssinus, Cope, loc. cit., p. 616.

PLASTOMENUS, Cope. U. S. Geol. Survey of Terrs. 1872, p. 617.

P. Thomasii, Cope, loc. cit., p. 618 ; *Trionyx Thomasii*, Pr. Am. Phil. Soc., 1872, p. 462.

P. trionychoides, Cope, loc. cit., 619 ; *Anostira trionychoides*, Pr. Am. Phil. Soc., 1872, p. 461.

P. multifoveatus, Cope, loc. cit., p. 619.

P. ædemius, Cope, loc. cit.; *Anostira ædemius*, Pr. Am. Phil. Soc., 1872, p. 461.

P. molopinus, Cope, loc. cit. p. 620 ; *Anostira molopinus*, Pr. Am. Phil. Soc., 1872, p. 461.

AMPHIBIA.

Incertæ sedis, 1, Cope. U. S. Geol. Survey, 1872, p. 633.

PISCES.

TELEOCEPHALI.

CLUPEA.

C. humilis, Leidy. Pr. Ac. Nat. Sc., 1856, p. 266; U. S. Geol. Survey, 1870, p. 369; do., 1871, p. 372; Cont. to Ext. Vert. Faun., p. 195.

C. alta, Leidy. Cont. to Ext. Vert. Faun., p. 196. (Cope refers these two species to *Diplomystus*)

C. pusilla, Cope. Pr. Am. Phil. Soc., 1870, p. 382; U. S. Survey, 1870, p. 429.

DIPLOMYSTUS, Cope. U. S. Geol. Survey Bulletin, vol. iii., No. 4, p. 808.

D. dentatus, Cope, loc. cit.
D. analis, Cope, loc. cit., p. 809.
D. pectorosus, Cope, loc. cit., p. 810.

OSTEOGLOSSUM.

O. encaustum, Cope. U. S. Geol. Survey, 1870, p. 430.

DAPEDOGLOSSUS, Cope. U. S. Geol. Survey; Bulletin, vol. iii., No. 4, p. 807.

D. testis, Cope, loc. cit.
D. æquipinnis, Cope, loc. cit., vol. iv., No. 1, p. 77.

ERISMATOPTERUS, Cope. U. S. Geol. Survey, 1870, p. 427.

E. Rickseckeri, Cope, loc. cit.
E. levatus, Cope, loc. cit., p. 428; *Cyprinodon levatus*, Pr. Am. Phil. Soc., 1870, p. 382.
E. Endlichi, Cope. U. S. Geol. Survey Bulletin, vol. iii., No. 4, p. 811.

AMPHIPLAGA, Cope. U. S. Geol Survey Bull., vol. iii., No. 4, p. 812.

A. brachyptera, Cope, loc. cit.

ASINEOPS, Cope. Pr. Am. Phil. Soc., 1870, p. 380; U. S. Geol. Survey, 1870, p. 425.

. squamifrons, Cope. Pr. Phil. Soc., p. 381; U. S. Survey, 1870, p. 426.
A. vividensis, Cope. U. S. Survey, 1870, p. 426.
A. pauciradiatus, Cope. U. S. Geol. Survey Bull., vol. iii., No. 4, p. 813.

MIOPLOSUS, Cope. U. S. Geol. Survey Bull., vol. iii., No. 4, p. 813.

M. abbreviatus, Cope, loc. cit.
M. labracoides, Cope, loc. cit., p. 814.
M. longus, Cope, loc. cit., p. 815.
M. Beanii, Cope, loc. cit., p. 816.

PRISCACARA, Cope. U. S. Geol. Survey Bull., vol. iii., No. 4, p. 816.

P. serrata, Cope, loc. cit,
P. cypha, Cope, loc. cit., p. 817.

P. liops, Cope, loc. cit., p. 818.
P. oxyprion, Cope, loc. cit., vol. iv., No. 1, p. 74.
P. Pealeii, Cope, loc. cit., p. 75.
P. clivosa, Cope, loc. cit., p. 76.

NEMATOGNATHI.

PIMELODUS.

P. antiquus, Leidy. Pr. Ac. Nat. Sc., 1873, p. 99 ; Cont. to Ext. Vert. Faun., p. 193.

PHAREODON, Leidy. Pr. Ac. Nat. Sc., 1873, p. 99 ; Cont. to Ext. Vert. Faun., p. 193.
P. acutus, Leidy, loc. cit.
P. sericeus, Cope, U. S. Geol. Survey, 1872, p. 638.

RHINEASTES, Cope. Pr. Am. Phil. Soc., 1872, p. 486; U. S. Geol. Survey, 1872, p. 638.
R. peltatus, Cope, loc. cit.
R. Smithii, Cope, loc. cit.
R. radulus, Cope. U. S. Geol. Survey, 1872, p. 639.
R. calvus, Cope, loc. cit., p. 640.
R. arcuatus, Cope, loc. cit., p. 641.

CYCLOGANOIDEI.

AMIA. *A. (Protamia) uintaensis*, Leidy. Pr. Ac. Nat. Sc., 1873, p. 98 ; Cont., p. 185.
A. (Protamia) media, Leidy, loc. cit.; Cont. to Ext. Vert. Faun., p. 108.
A (Protamia) gracilis, Leidy, loc. cit.
A. depressa, Marsh. Pr. Ac. Nat. Sc., 1871, p. 105.
A. Newberriana, Marsh, loc. cit.

HYPAMIA, Leidy. Pr. Ac. Nat. Sc., 1873, p. 98 ; Cont. to Ext. Vert. Faun., p. 189.
H. elegans, Leidy, loc. cit.

PAPPICHTHYS, Cope. U. S. Geol. Survey, 1872, p. 634.
P. sclerops, Cope, loc. cit., p. 635.
P. plicatus, Cope, loc. cit.
P. lævis, Cope, loc. cit., p. 636.
P. symphysis, Cope, loc. cit.
P. Corsonii, Cope, loc. cit.

RHOMBOGANOIDEI.

LEPIDOSTEUS. *L. atrox*, Leidy. Pr. Ac. Nat. Sc., 1873, p. 97 ; Cont. to Ext. Vert. Faun., p. 189.
L. simplex, Leidy. Pr. Ac. Nat. Sc., 1873, p. 98 ; Cont. to Ext. Vert. Faun., p. 191.
L. notabilis, Leidy, loc. cit.

L. glaber, Marsh. Pr. Ac. Nat. Sc., 1872, p. 103.
L. Whitneyi, Marsh, loc. cit. (These species are all referred to *Clastes* by Cope.)

CLASTES, Cope. U. S. Geol. Survey, 1872, p. 633.
C. anax, Cope, loc. cit.
C. cyliferus, Cope, loc. cit., p. 634.

www.ingramcontent.com/pod-product-compliance
Lightning Source LLC
Chambersburg PA
CBHW030350170426
43202CB00010B/1320